Freeze-Thaw
Durability of Concrete

Freeze-Thaw Durability of Concrete

Proceedings of the International Workshop in the Resistance of
Concrete to Scaling Due to Freezing in the Presence of De-icing Salts

Sainte-Foy, Québec, Canada

Papers from the International Workshop on
Freeze-Thaw and De-icing Resistance of Concrete

Lund, Sweden

Sponsored by CRIB, RILEM, CANMET, CPA,
the Network of Centres of Excellence on High Performance Concrete
and the Lund Institute of Technology

EDITED BY

J. Marchand
Centre de recherche interuniversitaire sur le béton
Université Laval
Québec
Canada

M. Pigeon
Centre de recherche interuniversitaire sur le béton
Université Laval
Québec
Canada

AND

M. Setzer
University of Essen
45117 Essen
Germany

CRC Press
Taylor & Francis Group
Boca Raton London New York

CRC Press is an imprint of the
Taylor & Francis Group, an **informa** business

A CHAPMAN & HALL BOOK

CRC Press
Taylor & Francis Group
6000 Broken Sound Parkway NW, Suite 300
Boca Raton, FL 33487-2742

First issued in paperback 2019

© 1997 RILEM 1997
CRC Press is an imprint of Taylor & Francis Group, an Informa business

No claim to original U.S. Government works

ISBN-13: 978-0-419-20000-0 (hbk)
ISBN-13: 978-0-367-86399-9 (pbk)

A catalogue record for this book is available from the British Library

Publisher's Note This book has been prepared from camera ready copy provided by the individual contributors in order to make the book available for the Conference.

Visit the Taylor & Francis Web site at
http://www.taylorandfrancis.com

and the CRC Press Web site at
http://www.crcpress.com

Contents

Preface

In North America as in several European countries, the frost-induced deterioration of concrete has been a matter of major concern for many years. Over the years, it has been clearly established that saturated concrete exposed to repeated freezing and thawing cycles can be affected by two types of deterioration: internal microcracking and surface scaling. It has also been shown that, in practice, each phenomenon can occur independently of the other.

In the past decades, laboratory experiments and field experience have clearly demonstrated that most concrete mixtures can be reliably protected against frost-induced microcracking by air entrainment. There also exists an overwhelming body of data that indicates that de-icing salts greatly amplify the surface scaling deterioration of concrete structures exposed to natural freezing and thawing cycles. Due to the widespread use of de-icing salts during winter roadway maintenance operations, de-icer salt scaling has become, in many Northern countries, one of the main causes of the premature deterioration of concrete structures.

Surface scaling is also one of the most complex problems facing the concrete construction industry. First of all, the freezing of water containing dissolved salts in a fine pore structure such as that of hardened cement paste is an extremely intricate phenomenon. The fact that scaling is essentially a surface phenomenon also adds to the complexity of the problem. Even when a structure is found to suffer from very severe scaling the deterioration is usually limited to the first millimetres of concrete near the surface. For a number of reasons, including the placement and curing conditions as well as drying and carbonation, the internal structure of this surface layer is often significantly different from that of the bulk of the concrete. Given the complexity of the problem, there still exists no reliable solution to fully protect normal concrete structures against de-icer salt scaling.

It is thus not surprising that a great deal of effort has been spent, over the past decades, towards understanding the basic mechanisms that control the frost durability, and more particularly the de-icer salt scaling resistance, of concrete. The publications presented in this book are part of this effort. They were prepared for discussion at meetings of the RILEM technical committee. This committee includes practising engineers and scientists involved in basic and industrial research on the frost durability of concrete. Most of the publications included in this volume were presented at one of the two workshops organised by the committee. The first workshop was held at the Lund Institute of Technology, Lund Sweden in June 1991. The second workshop was held at Laval University, Québec City, Canada in August 1993.

In the view of the editors, the publications included in this volume represent very significant contributions to the understanding of frost durability and the de-icer salt scaling resistance of concrete. Many of the publications are authored by some of the most respected experts in the field. The subjects addressed by the authors are diverse and cover many aspects of the frost durability of concrete. Special attention is paid to the crucial question of testing. It is hoped that this information will be useful not only to researchers, but also to practising engineers who are responsible for the design and construction of durable structures.

J. Marchand

M. Pigeon

M.J. Setzer

Acknowledgements

The editors would like to acknowledge the precious collaboration of all authors without whom the publication of this volume would not have been possible. The editors would also like to thank all RILEM TC 117 members for their assiduous participation to the work of the committee. The editors are grateful to MM. Rainer Auberg and Volker Hartmann who have successively acted as secretaries of the committee. The editors are also grateful to Professor Goran Fagerlund for organising the Lund workshop and Mrs Marthe Beauchamp for her participation in the organisation of the Québec City workshop. The various sponsors of the Lund and Québec City meetings are gratefully acknowledged for their financial support

Part I

Basic phenomena and
deterioration mechanisms

Action of frost and deicing chemicals – basic phenomena and testing

M.J. SETZER
University of Essen, 45117 Essen, Germany

Abstract
Three modifications of pore water are distinguished: Structured, prestructured and bulk. Water uptake by condensation and suction is outlined. A classification of pore sizes is given on this basis. The freeze thaw behaviour of the modifications is described. The influence of deicing chemicals is summarised. The basic prerequisites of a test procedure are derived from these considerations. The results with respect to precision, reproducibility, independence of testing institute and calibration to practical behaviour are shown.
Keywords: Frost action, pore water, pore classification, transport phenomena, basics of testing.

1. Introduction

Up to now in testing the freeze thaw durability it has been primarily attempted to copy outdoor conditions in the laboratory. However this is possible only in a very restricted manner since climatic variations are extremely high. The reaction of concrete to the changes in environmental conditions are only predictable if the special microstructural behaviour is taken into account. The freeze thaw and deicing salt test can only be optimised properly and developed into a reliable procedure if the factors affecting the results are analysed on the basis of the physical behaviour of pore water and the microstructure of concrete. It deviates from macroscopic experience significantly due to the highly dispersed microstructure. Especially the freezing point of pore water is depressed significantly depending on pore size and salt content. The amount of freezable pore water depends on the degree of water saturation, temperature, structure

Freeze-Thaw Durability of Concrete. Edited by J. Marchand, M. Pigeon and M. Setzer.
Published in 1997 by E & FN Spon, 2–6 Boundary Row, London SE1 8HN, UK.
ISBN 0 419 20000 2.

and the concentration of the dissolved deicing chemicals in concrete. If during the test one of these parameters is allowed to vary the test result will be unpredictable. Changes in surface tension lead to additional stress. The transport phenomena of water and of dissolved ions in the structure are complex.

Starting with the basic research on abnormal freezing by Powers, Brownyard and Helmuth [11,17,18,19] several authors have contributed to the present knowledge e.g. Everett [9], Fagerlund [10], Litvan [14], Selevold and Bager. We have also contributed some ideas on the description of the abnormal freezing of pore water by thermodynamic models [26,27,28]. The models were verified and refined by DSC measurements [1,2,3,29], low temperature measurement of elastic modulus [34,37] and thermal expansion [30]. Meanwhile we can distinguish several types of pore water modifications with characteristic freezing behaviour. This is outlined below. A relation to water transport and frost resistance is given.

Starting from this basic research work and the analysis and adoption of existing tests, a test procedure can be defined which meets the general demands of test procedures: Precision, reproducibility, independence of testing institute and correlation to practical behaviour. Simplicity of handling and economical price is fulfilled too.

2. Basic Considerations

The basic considerations are divided into two parts: 1. The classification of pores with respect to size. This comprises the water in these pores and its special behaviour. 2. The freezing behaviour of pore water.

2.1. Classification of Pores and of Pore Water

The physical properties of pore water are strongly affected by the interaction with the internal surfaces of the gel matrix. Depending on the distance from the internal surface, this interaction leads to water modifications which are fundamentally different from normal bulk water. On the basis of sorption data and DSC experiments we distinguish between three modifications of pore water [28]. The pore water modifications as well as the different mechanisms of water uptake and loss lead to a classification of pores. Both the definition of pore water modifications and the categorizing of pores lead to a better understanding of the frost phenomena and finally of testing freeze thaw and deicing salt resistance.

2.1.1. The Adsorbed Layer - Structured Surface Water

On the internal surfaces an adsorbed film of 1 to 3 molecular layers is formed. It is highly structured by the surface interaction and therefore substantially different from bulk water. Additionally this kind of water is found in very small gel pores with a hydraulic radius below 1 nanometer.

The hydraulic radius R_H is defined by the ratio of volume V by surface S. For cylindrical pores of length l the geometrical radius r is double the hydraulic radius.

$$R_H = \frac{V}{S} = \frac{1 r^2 \pi}{12 r \pi} = \frac{r}{2} \tag{1}$$

The properties of structured water e.g. its chemical potential, its influence on surface energy are described in more detail in [28] and in [36]. It comprises approximately 30 % of the total pore water in a completely water saturated concrete. Structured water stays unfrozen at usual temperatures.

2.1.2. Prestructured Condensed Water

Prestructured condensed water is condensed in gel pores between 50 and 98 % r.h. and contained in pores between 1 and 30 nm hydraulic radius. The condensation can be described by the Kelvin equation or by its modified form [27,28]. Since the degree of filling is dependent on relative humidity this condition is called "hygroscopic condition". The freezing point of prestructured water is depressed by both the change of surface interaction during freezing and a reduced molar entropy s_l of the prestructured water.

2.1.3. Bulk Water

It is contained in capillary pores with a hydraulic radius above 1 nm. However if concrete is exposed to air water is evaporating out of these pores. Bulk water is sucked up as a liquid due to the curved liquid air interface of capillary water. The curvature is determined by the surface interaction forces of the solid air interface - with adsorbed films - γ_{sa}, of the solid water interface γ_{sw} and the water air interface. The balance of these 3 forces determines the contact angle ϑ. Due the curved surface with the principal radii of curvature r_1 and r_2, a pressure difference p between liquid and air is created.

$$p = \gamma_{wa} \cos(\vartheta)\left(\frac{1}{r_1} + \frac{1}{r_2}\right) = \gamma_{wa} \cos(\vartheta)\left(\frac{1}{R_H}\right) \tag{2}$$

Here the hydraulic radius is defined by the radii of curvature. In case of a cylindrical pore $R_H = r/2$. ϑ is usually supposed to be 0° and therefore $\cos(\vartheta)$ to 1.

From this pressure a maximum suction height can be calculated (ρ is the density and g the gravitational acceleration)

$$h = \frac{p}{\rho g} = \frac{\gamma_{wa}}{\rho g} \cos(\vartheta)\left(\frac{1}{R_H}\right) \tag{3}$$

2.1.4. Pore Size Classification

Following Darcy's law the friction increases with the 4th power of the radius. Therefore the maximum height is reached after a period of time. In figure 1 the suction height is plotted as a function of the hydraulic radius of a pore for different suction times. It is seen that for pores with a radius above 30 μm the maximum height is reached within 1 minute, i.e. immediately. In this case the suction height is small. For hydraulic radii between 1 μm and 30 μm the maximum is reached between several minutes and several days. In smaller pores, however, the suction height is markedly below the maximum even after very long suction times.

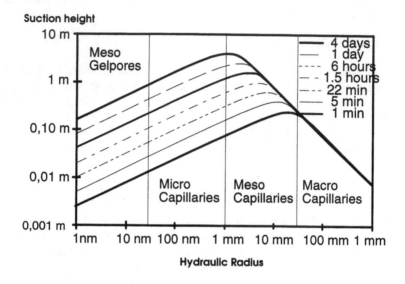

Figure 1. Suction height for different suction times between 1 minute to 4 days as function of hydraulic radius. The hydraulic radius of the pores is assumed to be constant.

Table 1. Pore size classification and pore water

Pore class	Upper R_H	Kind of water	Filled by
Micro gel pores	1 nm	structured	sorption (<50%r.h.)
Meso gel pore	30 nm	prestructured	vapour condensation (50 to 98 % r.h.)
Micro capillaries	1 μm	bulk	suction (no maximum height reached)
Meso capillaries	30 μm	bulk	suction (maximum height reached after minutes to some days)
Macro capillaries	1 mm	bulk	suction (maximum reached below 1 minute)

IUPAC defines pores below a radius (cylindrical pores) of 2 nm as micropores , between 2 nm and 50 nm as mesopores and above 50 nm as macropores . On the basis of this classification the distinction in table 1 can be made. However, here the hydraulic radius instead of the radius of a cylindrical pore is used.

In this classification a geometric series is applied. Water in gel pores is essentially different from bulk water. The micro gel pores are water filled above 50 %r.h. and thus, as a rule are full. The meso gel pores are more or less water filled depending on the relative humidity of the surrounding air. Capillaries are water filled only if liquid water is sucked up - macro capillaries instantaneously, meso capillaries over a period of minutes to days, micro capillaries, although the hydraulic pressure is extremely high, only partially and over small distances.

2.2. Freezing of Pore Water

Both a depression of freezing point and a supercooling of pore water are observed. The depression of freezing point is reproducible dependent on the microstructure and surface physics. Supercooling is a statistical effect with a random decrease in freezing point dependent on heterogeneous and homogeneous nucleation processes during the cooling period. In addition the different types of pore water are subjected to different kinds of freezing.

The structured adsorbed water near the internal surface is so strongly physically bound by surface interaction, that it stays unfrozen under usual conditions, even if the adjoining water is already ice. The unfrozen layer between ice and the internal surfaces of the solid decreases with decreasing temperature. Between -60 °C and -120° a relaxation process of the water at the interfaces is observed [37,38]. Nevertheless, the adsorbed layer is of high importance to the freezing behaviour of pore water. Due to the structured water film pore ice is not surrounded by air but by a medium similar to water. Thus the surface energy γ of the ice nuclei is much lower than for ice particles in the air (approximately by a factor of 10). In contrast to isolated ice particles of the same size pore ice crystals are stable and can be formed in situ.

The freezing point of pore water is depressed by the interaction with the internal surfaces of hardened cement paste. Due to this interaction a specific surface energy γ is generated. It is the integral over all interactions between two bulk phases. Before freezing it is the interaction between solid gel and bulk liquid water γ_{gl}, and after freezing the interaction between solid gel an ice γ_{gi}. The adsorbed structured film is located between gel and ice and, therefore, contained within the integration path. On freezing both the change of the surface interaction and the molar entropy is defining the freezing point and the heat of fusion. The depression of freezing point can be calculated from the differences of energies before and after freezing by a modified Clausius Clapeyron equation.

$$(s_i - s_l) = -(v_i \gamma_{gi} - v_l \gamma_{gl}) d\left(\frac{1}{R_H}\right) \tag{4}$$

(Here R_H is the hydraulic radius of the ice particles, T the absolute temperature, s the molar entropy, v the molar volume and γ the surface energy. The indices denote i ice, l liquid and g gel).

Near the freezing point $\Delta s = (s_i - s_l)$ can be approximated by the freezing enthalpy h = $T_0 * \Delta s$ and $(v_i \gamma_{gi} - v_i \gamma_{gi}) \approx v_i \gamma_{li}$. Integration of equation 1 and leads to the following approximation

$$\Delta T = (T_0 - T) = \frac{T_0 \gamma_{il} v_l}{hR_H} \tag{5}$$

A more precise calculation leads to the solutions by Brun et al. [4]. The surface tension is, also, temperature dependent:

$$\gamma_{si} = \{40,9 + 0,39(T - T_0)\} \quad mJ / m^2$$

A somewhat different approach is by Stockhausen [31]. The results are plotted in figure 2.

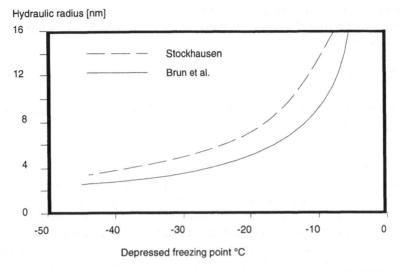

Figure 2.   Depression of freezing point due to restriction by pore radius [31].

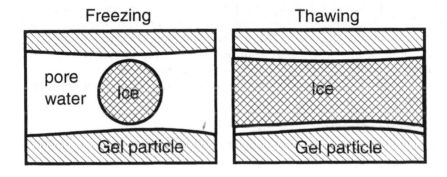

Figure 3.   Model of freezing and thawing pore ice.

In the calculations for figure 2 the changes of the molar entropies of bulk water and ice are taken into account more precisely. Stockhausen additionally used the modified chemical potentials for evaluation. However neither the surface free energy changes with temperature nor the changes of molar entropy due to prestructuring are taken into account. The interaction between water and ice and the entropies of bulk water are used instead. At present we are performing precise measurements in an attempt to fill this gap. At the time existing however we have only strong semiquantitative values.

In these models it has not been taken into account that on freezing the hydraulic radius of the ice nucleus R_{Hn} (figure 3 left) which fits into the pore is the appropriate value for the calculation of the depressed freezing point T_f, whereas for the melting temperature T_m it is the hydraulic radius of the pore ice particle R_{Hp} (figure 3 right) i.e. the hydraulic radius of the pore reduced by the adsorbed unfrozen film. The hydraulic radius of the ice nucleus R_{Hn} is much smaller than hydraulic radius of the ice particle R_{Hp} This is one reason why the freezing point is much more depressed than the thawing temperature. The hydraulic radius of the ice nucleus R_{Hn} is defined by the two principal radii of curvature r_1 and r_2:

$$\frac{1}{R_{Hn}} = \frac{1}{r_1} + \frac{1}{r_2} \tag{6}$$

Out of the two hydraulic radii a form factor f can be defined as a characteristic of the pore geometry.

$$f = R_{Hn}/R_{Hp} \tag{7}$$

In [1,3] we published scanning loops of differential scanning calorimetry, where we cooled down to a minimum temperature T_c and heated up again. The minimum temperature of a freeze-thawing cycle was increased from -70 °C in steps of 2 °C until 0 °C. The difference between two sequential cycles has been calculated.

From figure 4 it is seen firstly that the different cooling curves superimpose exactly whereas the melting curves are essentially different. The 3 most characteristic difference curves are plotted in the lower part of the figure. Using these scanning loops we were able to distinguish 3 characteristic phase transitions of pore water:

>From these findings it is evident that prestructured gel pore water does not freeze above -20 °C. Therefore in testing freeze thaw durability only macroscopic water sucked up by capillary pores can freeze. The prestructured water remains unfrozen. However it is transported to the ice particles when freezing sets in. This leads to an immediate expansion of the total concrete when freezing sets in due to expansion of the ice in the pores [30,31]. It is followed by a contraction due to shrinkage of the gel pore system.

2.3. Influence of Deicing Chemicals
In [3] we studied the influence of deicing chemicals on the freezing of pore water too.

Table 1.   Phase transitions of prestructured gel pore water

Transition		A	B	C
Freezing temperature	T_f °C	-39.4	-31.0	-23.7
Melting temperature	T_m °C	-24.6	-14.0	-9.2
Form factor	f	1.7	2.3	2.7

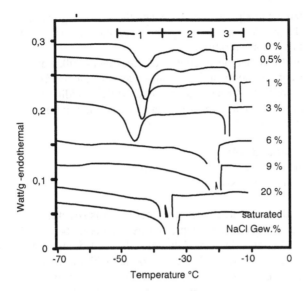

Figure 5. DSC diagrams of hardened Portland cement paste (OPC W/C= 0.5) prestored in sodium chloride solutions of different concentration [2,3].

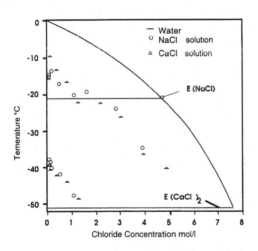

Figure 6. Depre of freezing point d chloride concentratior the superimposed surface interaction. line: normal bulk so. points in the higher r Freezing of capillary like - solution (transitic figure 5); points in the temperature region: Fr of prestructured transit (transition 1 of figure 5

From the DSC diagrams the following conclusions can be made:

- The influence of salt solutions on the depression of freezing point is superimposed on the influence of surface interaction. The superposition is linear and additive. With increasing concentration the freezing points shift to lower temperatures.
- The phase transitions of prestructured gel pore water are broadening and the areas under the peaks are diminishing with increasing salt concentration.

- The calculated freezing enthalpy is lowered. Since the freezing point is not affected by this lowered enthalpy. The surface energy between internal surface and pore ice should increase with salt concentration.
- There are some indications that the amount water freezing as bulk water increases.

A sodium chloride concentration of 3 % is most detrimental for freezing, since a significant amount of pore water freezes above -20°C while in addition the amount of unfrozen water is considerable. However it is seen too, that the minimum temperature can change this relation drastically.

2.4. Transport of Water and Salt
Water is taken up by the concrete structure in two ways:
1. adsorption and capillary condensation and
2. capillary suction.

As stated above the adsorbed and condensed water either is structured and does not freeze normally or is prestructured with a depressed freezing point below -20 °C.

Water which is sucked up is at least partially freezable depending on the minimum temperature. In addition deicing chemicals are transported with the water into the concrete structure. This combined transport of water and dissolved salts is much more efficient than the diffusion of salt into a specimen which is already water saturated [33].

Additionally the thermal expansion of liquid water is, apart from the anomaly of water, approximately 10 times larger than the thermal expansion of ice. Together with the freeze-thaw hysteresis this leads to a pumping effect. If there is an external water reservoir an additional water saturation is observed after a freeze thaw cycle. The water uptake is much larger than after usual capillary suction at room temperature.

Finally water transport takes place within the microstructure of concrete. It is generated by the changed surface interaction after ice is formed [27]. This is connected both with a sudden change of chemical potential and pressure differences. The change in chemical potential leads to a desiccation of small pores and a water uptake of the larger ice filled pores combined with shrinkage and swelling. The pressure differences lead similarly to contracting forces in small pores and either to smaller contracting forces or even expansion in the larger pores depending on boundary conditions.

Therefore transport phenomena during freeze thaw cycles lead to an additional water uptake and to micro structural changes and damage.

3.   Testing of the Freeze Thaw and Deicing Resistance

3.1.   Basic Considerations and Agreements in RILEM TC 117
The essentials for any testing should be:
- Precision i.e. small scattering of the results.
- Reproducibility.
- Independence from the testing institute.
- Correlation to practical behaviour.

From a economic point of view:
- Speed.
- Easy handling.

- Reasonable price of the equipment

According to the basic research some parameters are essential if testing of freeze thaw durability is to be reliable. They must be defined as precisely as possible, if the above mentioned main prerequisite of any testing procedure is to be met. They are:

1. The amount of freezable water play the dominant role. Therefore the degree of saturation must either be fixed as precisely as possible or be altered under extremely well controlled boundary conditions. This is in full agreement with the findings of Fagerlund [10].
2. The concentration of dissolved salts especially of deicing chemicals within the concrete must be adjusted with similar precision.
3. If the above mentioned points are fulfilled, the maximum ratio between frozen and unfrozen water depends extremely sensitive upon the minimum temperature reached. Therefore temperature control is one of the most critical test parameters..
4. Since transport both microscopic and macroscopic phenomena are generated by temperature changes the local and temporal temperature gradients must be given by precise boundary conditions, i.e. The heat flux must be defined.

In the first meeting of the RILEM TC the participants fully agreed that these basic points must kept in mind. In addition it was agreed that

1. the test should be uniaxial due to correlation to practical conditions both with respect to heat flux and water transport. Therefore the heat and water flux through all other sides apart from the surface to be attacked must be negligible.
2. the minimum temperature should be fixed to -20 °C.

3.2. The CDF Test - One Way of Realisation

By analyzing other test procedures RILEM TC tried to meet these demands with a (as far as possible) simple procedure and by adopting as many details of other methods as possible to come to comparable results. Finally, we defined CDF test (Capillary suction of Deicing chemicals and Freeze thaw test) [33].

From the Austrian standard ÖNORM B3303 [**Erreur! Source du renvoi introuvable.**] we used basically the temperature regime. However we found that two test cycles instead of one within 24 hours did not change the result significantly if the minimum temperature was kept constant sufficiently long (3 hours).

To meet a precise degree of saturation we used the physical phenomenon of capillary suction. This could be achieved by a predrying of the specimen under laboratory conditions (20 °C, 65 % r.h.). In this way all the freezable water is evaporated. For capillary suction the test area must be at the bottom of the specimen. To avoid side effects the lateral faces are sealed either with epoxy resin or with a bituminous band. The top of the specimen must be left unsealed to allow the air to escape during the suction process.

Using this procedure it is easy to fix the content of deicing salt, since it is sucked up with the water in a well defined way.

The most problematic point is certainly the temperature control. It must be kept in mind that during freezing (and inversely thawing) large amounts of heat of fusion must be removed. This heat is significantly larger than the heat capacity of the concrete matrix. The heat of fusion of the water or solution reservoir adjacent to the surface of attack is not negligible - especially the thickness of this layer.

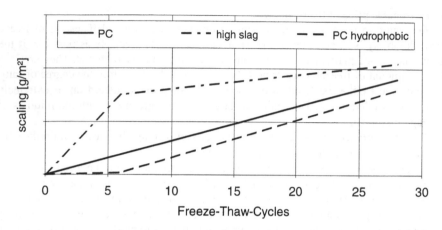

Figure 7.   CDF test: Scaling of a typical Portland cement concrete, a high slag cement concrete and a concrete with hydrophobic surface.

Usually heat is transported in climatic chambers by circulating air. However the heat capacity of air is too small to transport remarkable amounts of heat within a short time. On the other hand non convecting air is an excellent thermal isolator. Therefore we used the insulating capacity of air. For heat transport we used a liquid bath. For this purpose the tested area must be face down during the freeze thaw cycle too. This arrangement has several additional advantages. The capillary suction is defined. The - remarkably high - heat capacity of the liquid layer on the concrete surface is limited by its thickness i.e. the distance between the concrete surface and the bottom of the container. There is no insulating air layer between the liquid and the cooling medium as in air cooled machines with the test area at the top of the specimen. Finally, the arrangement can be used to remove the deteriorated loose material in a defined way from the surface with an ultrasonic bath. The temperature can be defined with high precision (± 0.3 °C). The thermal gradient is uniaxial. An additional lateral thermal isolation is unnecessary since this is already provided by the static air.

By these arrangements the above mentioned parameters proved to be excellently defined.

At the moment the amount of deteriorated material is used as criterion.

Additionally, we found that after a few temperature cycles (≈ 6) the scaling was proportional to the number of cycles. The predictability is excellent. The deviation during the first cycles is either due to the scaling of the carbonated layer or to the retardation of protective sealing or of a hydrophobic agent.

Meanwhile we are able to verify the results by more than 400 tested mixes and 4 round robin tests.

3.3. Precision of Test

Already in [21,23] is noted that the standard deviation s or the coefficient of variation v are correlated to the mean value of scaling $\overline{w}$ by a power function at least for CDF test and the test according to Austrian ÖNORM 3303:

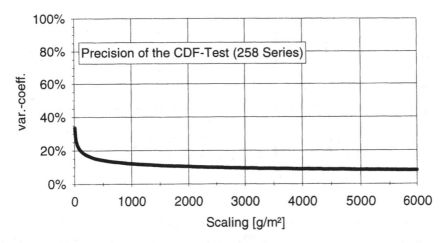

Figure 8.   Correlation of coefficient of variation and mean scaling for CDF test on the basis of 258 tests

$$v = a\overline{w}^n \tag{8}$$

Here a and n are constants depending on the test procedure. Meanwhile this relationship proved to be valid for other tests too e. g. the cube test, the Boras test.

The most essential point in this relationship is that the exponent n must be smaller than zero. Otherwise an increase of test cycles and of scaling would lead to a decrease of precision. In a double logarithmic scale the power function is linear.

In figure 8 this correlation is plotted for CDF test. On the basis of this graph two essential point for a test procedure can be defined: The level of an acceptance criterion and the minimum number of test specimens. Meanwhile, the precision of CDF test has been assessed following ISO 5725 in several round robin tests [25].

The coefficient of variation v for repeatability, reproducibility and between laboratory scattering is dependent from the scaled mass m following the equation:

$$v = v_0 \left(\frac{m}{m_0} \right)^d \tag{9}$$

Herein $m_0 = 1500$ g/m². This value can be used as acceptance limit. v_0 and d are given by the following table:

Table 2.   Precision of CDF test [25] assessed following ISO 5725 . The repeatability gives the precision in one test laboratory. The reproducibility gives the precision which is reached comparing the results of different laboratories of one mix.

	Repeatability	Between laboratory	Reproducibility
d	-0.33	-0.26	-0.29
v0	10.4 %	14.0 %	17.5 %

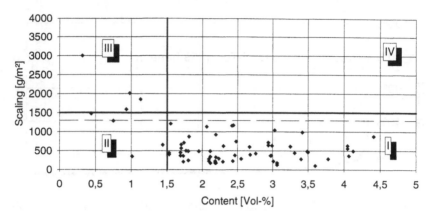

Figure 9. Scaling - mean value - of concrete as a function of the content of artificially entrained air void below 300 μm (A300 content). The critical value of A300 content 1.5 % and the CDF criterion - 1500 g/m² are marked as solid lines, the safety margin as outlined below as dotted lines. The Latin numbers refer to the distinction of areas below.

3.4. Calibration

3.4.1. Required Number of Freeze Thaw Cycles

The acceptance criterion must define the quality of a concrete, which marks the border between acceptable and not acceptable. The distinction must be possible with sufficient precision and confidence. A confidence level of 95 % should be reached. Obviously the confidence level increases if the coefficient of variation decreases. From statistical considerations [25] we wanted a value near 10 % for good testing. As stated above the coefficient of variation decreases with the mean value of scaling. By evaluating more than 350 tests we can state that a coefficient of variation of 11 %, i.e. a sufficiently high precision, is reached at a mean value of scaling of 1500 g/m². For the CDF test with its constant scaling rate it will be shown below that this level is reached after 28 cycles.

Therefore we can define the required number of cycles as 28 for the CDF test.

3.4.2. Acceptance Criterion

It is almost impossible to calibrate a test of freeze thaw and deicing salt resistance by simply observing the real performance of concrete under practical conditions since this is a very long and highly inaccurate procedure. Nevertheless a calibration is possible. Rules for mixes which are sufficiently resistant against freeze thaw and deicing salt attack are defined in the standards. On testing these mixes a critical scaling for a certain test procedure can be found. We tested more than 80 different compositions. Additionally we measured the characteristics of entrained air voids. While the total air

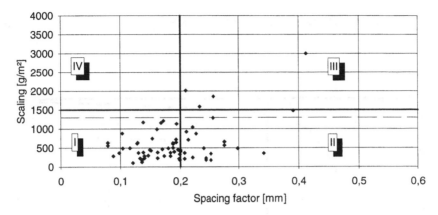

Figure 10. Scaling - mean value - of concrete's dependent from spacing factor - the critical value of 0.2 mm is marked. Other details as in figure 9.

content does not allow a reasonable distinction, the content of voids below 300 μm (A300) and the spacing factor have acceptable values.

In figures 9 and 10 four different areas can be distinguished if we define three border lines: a critical mean scaling following CDF-test w_{cr}= 1500 g/m²; a critical air void content $A300_{cr}$= 1.5 vol.% and a critical spacing factor a_{cr}= 0.2 mm.

I. $w<w_{cr}$ and $A300>A300_{cr}$ in figure 9 or $a<a_{cr}$ in figure 10: In this area following from both the CDF test and A300 content or spacing factor a sufficient durability is predicted.

II. $w<w_{cr}$ and $A300<A300_{cr}$ in figure 9 or $a>a_{cr}$ in figure 10: In this area the air void parameters are not sufficient whereas the scaling is still acceptable. For these concrete apparently the traditional parameters do not predict the resistance adequately.

III. $w>w_{cr}$ and $A300<A300_{cr}$ in figure 9 or $a>a_{cr}$ in figure 10: In this area following from both CDF test and A300 content or spacing factor durability not sufficient.

IV. $w>w_{cr}$ and $A300>A300_{cr}$ in figure 9 or $a<a_{cr}$ in figure 10: Results in this field are critical, since in a CDF test the concrete's would fail whereas the air void parameters are sufficient. However any point we found in this field is related to a concrete with a type of cement which is not allowed to be used in concrete durable against freeze thaw and deicing salt attack.

Therefore we found a well suited calibration for the test procedure. It can also be seen that the CDF criterion of a upper limit for the mean scaling after 28 freeze thaw cycles at $\overline{w}_{cr}$=1500 g/m² is sufficiently sharp too.

For a acceptance criterion a upper 5 % fractile is the appropriate value. However all the tests were made with 5 specimens. Therefore a fractile cannot be given for every single test. However out of the total number of tests a coefficient of variation of 11 % at a mean scaling of 1500 g/m² can be assumed with sufficient certainty. (We found 11 % out of 254 data sets). Using standard statistical methods a critical upper 5 % fractile $w_{cr5\%}$ can be defined as acceptance criterion:

$$w_{cr5\%} = (1 + 1.64\,v)w_{cr} = 1.2 * 1500\,\frac{g}{m^2} = 1800\,\frac{g}{m^2} \qquad (9)$$

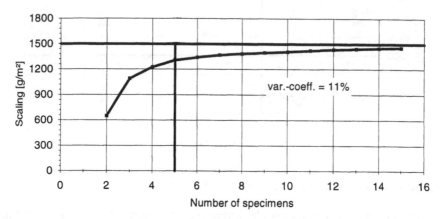

Figure 11. Maximum mean value of scaling for tests with different numbers of specimens to assure that the mean scaling of 1500 g/m² is not exceeded with a confidence level of 95 %. The variation coefficient is assumed to be 11 %.

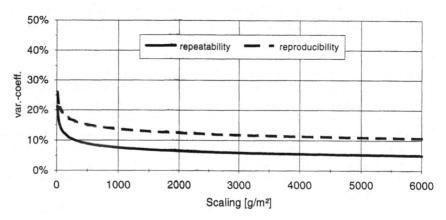

Figure 12. Coefficient of variation as function of scaling for the repeatability and reproducibility for CDF test

3.5. Procedure for a Single Test

On this basis the procedure for a single test with 5 specimens can be specified. A confidence level of 95 % shall be reached. Here a safety margin must be used which depends on the coefficient of variation and the number of specimens. The relevant parameter is coefficient of variation measured by the testing institute. It can be assumed by experience from former tests or calculated from the test results.

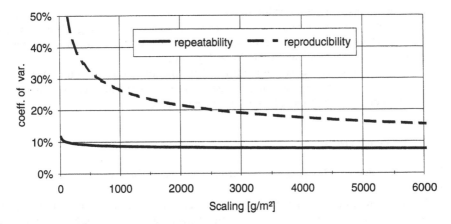

Figure 13. Coefficient of variation of the round robin CDF tests by international testing institutes as function of scaling.

As example a value of 11 % is assumed. Based on these assumptions the mean value for of 5 specimen must not exceed 1300 g/m^2 after 28 freeze thaw cycles n a single test. (See figure 11.) No single value must exceed 1800 g/m^2. Even on this basis the calibration is reliable.

For continuous testing with many specimens the 5% fractile criterion at 1800 g/m^2 after 28 freeze thaw cycles can be used for performance testing.

The mean value must not exceed 1500 g/m^2 in any case.

4. Conclusions

Up to now the testing of freeze thaw and deicing durability was sufficiently neither precise nor reproducible nor independent from the testing institute. Only institutes with highly experienced staff could distinguish between good and bad concrete based on their local experience. It has not been ascertained whether this singular know how could be transformed to a generalized procedure. On the other hand round robin tests showed high discrepancies even between experienced laboratories. The concrete in these tests were as a rule so distinctly different in mix design that durability testing was not necessary from a practical point of view.

However analysing the basic research work on the abnormal freezing of pore water in concrete allows definition of the relevant test parameters. If these parameters are sufficiently well defined it can be proved - e.g. with CDF test - that for a test procedure of freeze thaw and deicing durability the following prerequisites of any test can be reached:

- Precision with a coefficient of variation at below 11 %;
- Reproducibility with a coefficient of variation at below 10 %;
- Comparability between international testing institutes with a coefficient of variation at below 25 %;
- Calibration on practical behaviour with sufficient sharp distinction.

There are still some questions left open:

- The minimum temperature is extremely essential for testing. However the agreement to fix this temperature at -20 °C is very costly from an experimental point of view. A modification should be discussed.
- The temperature regime can be modified. This should be discussed too. A constant cooling rate is costly since it needs high cooling power near freezing point. The constant cooling rate cannot be justified by natural attack.

5. Acknowledgement

The author thanks Dr. Beddoe for discussions and Mr. Auberg for statistical evaluations.

The basic research was supported by the 'Deutsche Forschungsgemeinschaft', the development of the CDF test by the 'Bundesverband der Deutschen Beton und Fertigteilhersteller' and by the 'Forschungsgemeinschaft Transportbeton'.

6. Literature

1. Beddoe, R., Setzer, M.J. (1990) Phase transformations of water in hardened cement paste. A low temperature DSC investigation. *Cem. Concr. Res.*, Vol. 20, 236-242

2. Beddoe, R.E., Setzer, M.J. (1988) A low temperature DSC investigation of hardened cement paste subjected to chloride action. *Cem. Concr. Res.*, Vol. 18, p. 249

3. Beddoe, R.E.; Setzer, M.J. (1990) Änderung der Zementsteinstruktur durch Chlorideinwirkung (Change of structure of hardened cement paste by chloride action). *Forschungsberichte Fachbereich Bauwesen,* Vol. 49, University Essen, Essen,

4. Brun, M., Lallemand, A., Quisson, J.-F., Eyraud, C. (1977) A new model for the simultaneous determination of the size and shape of pores. *Thermochimica Acta,* Vol. 21, p. 59

5. EMPA   (2. Aufl. 1987 bzw. 3. Aufl. 1989*) Versuchsrichtlinie Prüfung Nr. 6: Porenkennwerte* SIA 162/1 Ziffer 3 06;

6. EMPA (2. Aufl. 1987 bzw. 3. Aufl. 1989) *Versuchsrichtlinie Prüfung Nr. 9: Frost-Tausalzverhalten.* SIA 162/1 Ziffer 3 09;

7. EMPA: (1989) *Versuchsrichtlinie: Kritischer Sättigungsgrad,*

8. EMPA (1989) *Versuchsrichtlinie: Schnellporosität,*

9. Everett, D.H. (1961) *Thermodynamics of frost damage to porous solids. Trans. Faraday Soc.,* Vol. 57, p. 1541

10. Fagerlund, G. (1977) The critical degree of saturation method of assessing the freeze thaw resistance of concrete. *Materials and Structures,* Vol. 10, p. 217-229

11. Helmuth, R.A. (1960) Capillary size restrictions on ice formation in hardened cement pastes. *4th Intern. Symp. on chemistry of Cement*, Washington, p. 855

12. Int. Kolloquium: (1980) *Frostbeständigkeit von Beton.* Wien, Mitt. VÖZ Heft 33

13. Jessberger, H.L. (1971) Bodenfrost. *Straßenbau und Straßenverkehrstechnik*, Heft 125
14. Litvan, G.G. (1975) Phase transitions of adsorbates : IV. Effect of deicing agents on the freezing of cement paste. *I. Amer. Ceramic Soc.*, Vol. 58, p. 26-30
15. ÖNORM B 3303 (3.83) *Betonprüfung*
16. ÖNORM B 3306, (9.82) *Prüfung der Frost-Tausalz-Beständigkeit von vorgefertigten Betonerzeugnissen.*
17. Powers, T.C., Brownyard, T.L. (1945) Studies of the physical properties of hardened portland cement paste : The freezing of water in hardened portland cement paste. *J. Amer. Concrete Inst. Proc.* 43 933-969
18. Powers, T.C., Helmuth, R.A. (1953) Theory of volume changes in hardened portland cement paste during freezing. *Proc. Highway Res. Board,* Vol. 32 p. 285-295
19. Powers1, T.C. (1945) A working hypothesis for further studies of frost resistance of concrete. *J. Amer. Concrete Inst. Proc.*, Vol. 41, p. 285-295
20. RILEM 4 CDC 1977 (1977) Methods of carrying out and reporting freeze thaw tests on concrete with deicing chemicals. *Materials & Structures*, Vol. 10 p. 213-215
21. Setzer, M. J. (1990) Prüfung des Frost-Tausalz-Widerstand von Betonwaren. *Forschungsberichte Fachbereich Bauwesen*, Vol. 49, University Essen, Essen,
22. Setzer, M.J., Hartmann, V. (1991) CDF-Test - Test procedure (with engl. translation). *Betonwerk und Fertigteiltechnik*, Vol. 57, Heft 9, p. 83-86
23. Setzer, M.J., Schrage, I. (1991) Testing for frost deicing salt resistance of concrete paving blocks. *Betonwerk und Fertigteiltechnik*, Vol. 57, Heft 8, p. 58-69
24. RILEM TC 117 (1995) CF and CDF Test - Test Method for the Freeze-Thaw Resistance of Concrete -Tests with Water (CF) or with Sodium Chloride Solution (CDF) : Part 1: Description of the test procedure, Part 2: Evaluation and precision of scaling in testing with sodium chloride solution Draft of a RILEM recommendation. *Materials and Structures.* Vol. 28, p. 175-182
25. Setzer, M.J., Auberg, R. (1995) Freeze Thaw and Deicing Salt Resistance of Concrete Testing by CDF Method - CDF Resistance Limit and Evaluation of Precision. *Materials and Structures*, Vol. 28, p. 16-31
26. Setzer, M.J. (1988) An improved model of hardened cement paste by low temperature and sorption studies. *Proc. Engineering Foundation Conf. Cement manufacture and use*, Potosi
27. Setzer, M.J. (1977) Einfluß des Wassergehalts auf die Eigenschaften des erhärteten Betons. *DAfStb* Vol. 280, Berlin, W. Ernst u. Sohn p. 44
28. Setzer, M.J. (1990) Interaction of water with hardened cement paste. In: S. Mindess ed.: *Advances in cementitious materials*. Ceramic Transactions Vol. 16, Am. Cer. Soc. Westerville, Ohio, p.415-439
29. Stockhausen, N., Dorner, H., Zech, B., Setzer, M.J. (1979)Untersuchung von Gefriervorgängen in Zementstein mit Hilfe der DTA. *Cem. Concr. Res.* Vol. 9, p. 783-794
30. Stockhausen, N., Setzer, M.J. (1980) Anomalien der thermischen Ausdehnung und Gefriervorgänge in Zementstein. *Tonindustrie Zeitung.* Vol. 2/104, p. 83-88
31. Stockhausen, N. (1981) *Die Dilatation hochporöser Festkörper bei Wasseraufnahme und Eisbildung.* PhD thesis, TU München

32. TFB(Schweiz) (1986) Prüfung von Festbeton auf Frost- und Frost-Tausalz-Beständigkeit. *Cementbulletin* Vol. 54, p. 10

33. Volkwein, A. (1991) Untersuchungen über das Eindringen von Wasser und Chlorid in Beton. *Berichte aus dem Baustoffinstitut* Vol. 1/91, TU München, München,

34. Zech, B., Setzer, M.J. (1988) The dynamic elastic modulus of hardened cement paste; Part I: A new statistical model - water and ice filled pores. *Materials & Structures.* Vol. 21, p. 323; (1989) Part II: Ice formation, drying and pore size distribution. *Materials & Structures* Vol. 22, p.125

35. Zech, B. (1981) *Gefrierverhalten des Wassers im Beton.* PhD thesis. TU München,

36. Setzer, M.J. (1996) Thermodynamic description of pore water. *Schriftenreihe des Fachbereichs Bauwesen,* Universität GH Essen, Vol. 64, Essen

37. Xu, X (1995) Tieftemperaturverhalten der Porenlösung in hochporösen Werkstoffen. PhD thesis University of Essen,

38. Xu, X.; Setzer, M.J. Damping maximum of hardened cement paste at -90 °C; A mechanical relaxation process. *Advances Cement Based Materials* to be published

2

On the service life of concrete exposed to frost action

G. FAGERLUND
Lund Institute of Technology, Lund, Sweden

Abstract
A theory is presented for the estimation of the service life of concrete exposed to frost action. It is based on a separation of the frost resistance problem in two parts:

Part 1, that is only a function of the material itself, and which is expressed in terms of a critical water content, or a degree of saturation, which is a "fracture value" being almost independent of the external conditions (except for the lowest temperature and the internal salt concentration.) The critical degree of saturation is a consequence of the existence of a critical distance of water transfer during freezing. Some experimental data for the critical distance are given.

Part 2, that is a function of the wetness of the environment, and which is expressed in terms of a time function of the capillary water uptake, and the long term water absorption in the air-pore system. Probably, it is also a function of the salt concentration outside and inside the concrete. It is shown in the paper, that the long term water absorption in pure water can often be described and extrapolated by simple time functions.
Keywords: Frost resistance, salt scaling, service life, capillarity, water absorption, spacing factor, air-pore structure.

1 Destruction mechanisms - the critical size

Water confined in the pore system of a porous material, when freezing, exposes the pore walls to stresses, which might, in some cases, cause severe damage to the material. The explanation closest at hand is the 9% increase in volume of water when it is transformed into ice. Excess water is expelled from the freezing site to an adjacent air-filled space, that is large enough to accomodate the water. During this flow, which occurs in a very narrow and partly ice-filled pore system, the pore walls are exposed to stresses, that are often referred to as hydraulic pressure. The mechanism was analyzed theoretically by Powers [1].

Another destruction mechanism, which theoretically should be of particularly great importance in very dense materials (low w/c-ratio), and/or when freezing occurs in the presence of de-icing salts, is analogous with the mechanism, that causes frost heave in the ground. Ice bodies, which have already been formed in coarser pores are, due to differentials in free energy between ice and water, able to attract unfrozen water from finer capillary pores and gel pores. Thus, there will be a water transfer towards the

Freeze-Thaw Durability of Concrete. Edited by J. Marchand, M. Pigeon and M. Setzer.
Published in 1997 by E & FN Spon, 2–6 Boundary Row, London SE1 8HN, UK.
ISBN 0 419 20000 2.

freezing site. The ice body will grow, and it will thereby expose the pore walls to pressure. The free energy of the ice body will, therefore, increase, at the same time as the free energy of the unfrozen water decreases due to the internal desiccation caused by the water flow. The growth of the ice body will not cease until its free energy is equal to that of the unfrozen water. Considerable pressure, high enough to seriously damage the material, could be built up. The pressure ought to be enhanced when the pore system contains salts; [2], [3]. The mechanism has been treated theoretically by numerous scientists. It was first applied to concrete by Powers and Helmuth; [4].

The relative importance of these two damage mechanisms has not been fully clarified. It is quite clear, however, that the second mechanism, in order to be significant, requires a large fraction of unfrozen water also at rather low temperatures. Therefore, it ought to be most pronounced in materials with low w/c-ratio.

There are also other destruction mechanisms suggested, or modifications of the two mechanisms described above. Such theories have been put forward by Everett [5], Haynes [6], Dunn and Hudec [7], Setzer [8], Litvan [9], and others.

One can theoretically show, that the hydraulic pressure mechanism, as well as the "growing ice body" mechanism, predicts that the destructive forces increase with increasing distance between the freezing site and the nearest air-filled space; [1], [4]. The concrete will be damaged in areas where this distance exceeds a certain critical value. This critical distance will adopt an individual value for each individual concrete and specimen geometry. One such critical distance is the critical thickness being the thickest possible completely water saturated plate of the actual concrete. Another critical distance is the critical shell thickness (or the critical Powers spacing factor) being the thickest possible water saturated cement paste shell surrounding an air-filled spherical cavity having an impermeable outer surface. There is also a critical sphere etc. Simple relations exist between the different sizes; if one of them is known, any other can be calculated, [10].

The relation between the critical thickness D_{CR} and the critical spacing factor L_{CR} is:

$$D_{CR} = 2 \cdot L_{CR}[2 \cdot \alpha \cdot L_{CR}/9+1]^{1/2} \tag{1}$$

where α is the specific area of the air-void which is enclosed by the shell. Thus, for a given concrete, the relation between L_{CR} and D_{CR} depends on the specific area of the air-pore system. (Note; strictly speaking, Eqn (1) is only valid when destruction is caused by the hydraulic pressure mechanism.)

2 The fictitious and the true spacing factor

Many investigators have proven experimentally the existence of critical distances of concrete; e.g. Ivey and Torrans [11], and Bonzel and Siebel [12]. The values seem to depend on the environmental conditions; the critical Powers spacing factor $(L_o)_{CR}$ is often claimed to be about 0.25 mm for freezing in pure water, and 0.16 to 0.20 mm for freezing in 3 % NaCl-solution. For very dense concretes, the values seem to be higher; [13]. The critical spacing factors are always determined by comparing the results of optical studies of the air-pore structure with results of freeze/thaw experiments of companion specimens. The values $(L_o)_{CR}$ are therefore based on the assumption that all spherical air-pores observed in the specimen are actually air-filled during the freeze/thaw test. This is normally not the case, however. Therefore, the values $(L_o)_{CR}$ determined in this way are fictitious, and considerably smaller than the true values L_{CR} based on the fraction of the air-pore system, that is actually air-filled during the freeze/thaw test. The significance of the two different spacing factors is discussed below.

The spacing factor L of a system of spherical pores in a material matrix is calculated by the Powers equation.

$$L = (3/\alpha) \cdot [1.4 \cdot (V_p/a + 1)^{1/3} - 1] \tag{2}$$

where α is the specific area of the part of the pore system the spacing of which is of interest, a is the volume of this pore system and V_p is the material volume within which the actual pore system is located, exclusive of the pores themselves. Normally, in a concrete, V_p is the volume of the cement paste phase inclusive of fine aggregate interfering with the pores.

3 Direct measurements of the critical true spacing factor

Some attempts have been made to experimentally determine the order of size of the critical true spacing factor of cement paste; [14], [16]. In [16], the w/c-ratio varied between 0.3 and 0.8. 3-year old water-cured, and completely saturated air-free specimens, were pre-dried at +50°C, and were then re-saturated by vacuum treatment. They were immersed, either in pure water, or in 3 % NaCl-solution, whereupon they were freeze/thaw tested between room temperature and -20 °C with a rate of temperature lowering of 2,5 °C/h in the interval 0°C to -10°C. The duration of each cycle was 16 hours of freezing, and 8 hours of thawing. Since the specimens contained no air, they were fragmented already during the first few cycles. An increasing number of cycles did not decrease the size of the fragments. This size is a measure of the critical thickness. The fragmented specimen was subjected to a sieve analysis, on basis of which the critical thickness could be estimated. The results are shown in Fig.1. The critical thickness is almost independent of the w/c-ratio, and is lowest for freezing in pure water. The following mean values are valid for w/c > 0,40 to 0,45. L_{CR} has been calculated by Eqn (1) assuming that the specific area is 15 mm^{-1}.

Freeze/thaw in pure water: D_{CR}=1.2 mm; L_{CR}=0.40 mm
Freeze/thaw in 3% NaCl-solution: D_{CR}=1.8 mm; L_{CR}=0.54 mm

Frost damage should, therefore, occur when the spacing factor between air-filled pores exceeds about 0.40 mm for freezing in pure water, and about 0.54 mm for freezing in salt solution. This might seem remarkable since, as mentioned above, the critical fictitious spacing factor is lower for freezing in salt solution. Also, in practice, freezing in salt water produces considerably more severe damage. The explanation closest at hand is, that the salt gives rise to a higher moisture content in the concrete surface. The positive effect of the larger critical size is, therefore, counteracted by an increased water absorption in the air-pore system; see paragraph 9. Another remarkable observation is, that the critical spacing factors are much higher than the fictitious values (about 0.18 mm and about 0.25 mm for freezing in salt and pure water.)

4 Indirect measurements of the critical true spacing factor

The critical true spacing factor (hereafter called the critical spacing factor) can also be calculated theoretically, provided the so called critical degree of saturation S_{CR} and the air-pore distribution curve are known. The mathematical procedure is presented in [15]. Calculations made indicate a rather good agreement with direct measurements of L_{CR}.

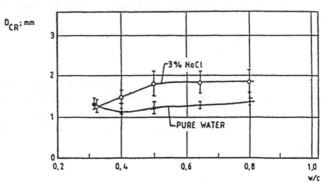

Fig. 1. Experimental determinations of the true critical thickness of OPC-pastes which have undergone a drying/wetting cycle; [16].

In [16] a calculation is made for a concrete with a w/c-ratio of 0.54, a total air content a_o of 7.1 %, and an experimentally determined S_{CR}-value of 0.80 (the same concrete as in Fig. 3.) The air-pore distribution curve was determined by means of the "manual" linear traverse method. The calculated L_{CR}-value is between 0.37 mm and 0.42 mm, which can be compared with the experimentally determined mean value 0.40 mm; Fig. 1. The different values 0.37 and 0.42 mm are based on different definitions of the cement paste volume; 0.42 implies that all aggregate particles smaller than 0.5 mm are included in the "cement paste" phase; 0.37 implies that no aggregate particles are included. According to an analysis performed in [17], a certain fraction of the finer aggregate ought to be included when calculating the spacing factor. The reason is that the inter-particle spacing of this aggregate fraction is of the same order of size as the air-pores.

Some other examples are shown in Fig. 2. The calculations are based on measurements of the S_{CR}-values, and the air-pore distribution curves, of 8 concretes containing cement with different amounts of ground granulated blast furnace slag; [18]. Two different concretes with different air contents were tested for each cement type. The two L_{CR}-values differ somewhat. One reason for this is that the the real air-pore distributions were not utilized, but merely the standard distribution according to Eqn (3), adjusted to the measured curve. The agreement between the theoretical and the real distributions is, however, not perfect.

$$f(r) = n \cdot \ln b/b^r \tag{3}$$

where r is the pore radius, and f(r) is the frequency function of pore radii. n and b are constants adjusted to the measured distribution. The absolute values of the calculated L_{CR}-values are a bit smaller than those found by direct freezing; see paragraph 3.

5 The critical degree of saturation

The existence of a critical size implies that there exists a critical water absorption in the air-pore system, and that there exists a critical degree of saturation of the concrete as a whole; [15]. For a given value of L_{CR}, the critical air-pore absorption and, therefore, the S_{CR}-value will be different for different air-pore distribution curves and air contents. Thus, the S_{CR}-value is individual for each concrete, despite the fact that the

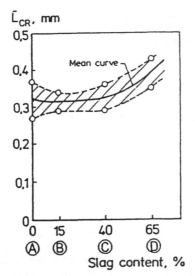

Fig. 2. The true critical spacing factor of slag cement concretes. The values are based on experimentally determined S_{CR}-values and air-pore distributions; [18].

value of L_{CR} is the same. This was first pointed out by Warris [19]. Degree of saturation is defined:

$$S = W_e/\varepsilon \ \text{[volume/volume]} \tag{4}$$

where W_e is the total evaporable water volume in the concrete and ε is its total pore volume (porosity) including all compaction pores, air-pores and aggregate pores.

In almost all cases, the occurrence of the moisture level S_{CR} implies, that a certain fraction of the air-pore system is water-filled. This is the reason why the experimentally determined true L_{CR}-values, presented above, are considerably larger than the fictitious $(L_0)_{CR}$-values (0.25 mm and 0.16 to 0.20 mm) on which a certain agreement has existed for a long time. These "traditional values" are valid for the entire air-pore system and imply that no part of this is water-filled during freeze/thaw. Therefore, in determining $(L_0)_{CR}$ by the Powers equation (2), the specific area of the air-pore system has always been assumed to be α_0, which is valid for the "dry" air pore system, although in reality, one should use a much smaller value α_r valid for that part of the air-pore system, that is actually air-filled; see Fig. 12.

The smallest air pores are water-filled at first, [20], and since they are numerous, they contribute very much to the value of α_0. Therefore, the difference between α_0 and α_r is often big. Besides, in calculating the value of the fictituous spacing factor $(L_0)_{CR}$, the total air content a_0 is used, and not the real volume a_r, that is actually air-filled. The real spacing factor L_r between air-filled pores in a specimen, that is freeze-tested, should of course be calculated by Eqn (2) using the values α_r and a_r, which are valid for the actual specimem considering its actual water absorption.

The S_{CR}-value can be calculated theoretically when the air content a_0, the radius distribution of the air pore system $f(r)$, and the true value L_{CR} are known; Eqn (5). The method of calculation is shown in detail in [15]. The only assumption made is that a smaller air pore is water-filled before a larger pore, which is a very reasonable assumption from a thermodynamical point of view.

$$S_{CR} = f[a_0;f(r);L_{CR}] \tag{5}$$

Inversely, the L_{CR}-value can, as said above, be calculated when the S_{CR}-value, the total air content, and the pore size distribution are known.

$$L_{CR} = g[a_0;f(r);S_{CR}] \tag{6}$$

6 Experimental determination of S_{CR}

The S_{CR}-value can be determined experimentally by methods described in [21]. A number of specimens are adjusted to individual degrees of saturation by drying from a vacuum-saturated condition, or by absorbing water after vacuum treatment of the dry specimen to a certain residual pressure. They are immediately sealed from moisture gain or loss, and are freeze/thaw-tested during one cycle only, or during a few cycles. Damage is measured by means of expansion or dynamic E-modulus. A plot of damage versus degree of saturation reveals the S_{CR}-value. Examples of determinations of S_{CR} of concrete and other materials can be found in many reports i.a. [17], [18], [21], [22], [23], [24], [25], [26], [27]. One example is seen in Fig. 3.

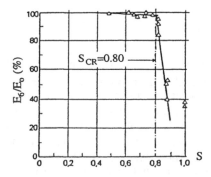

Fig. 3. Example of an experimental determination of the critical degree of saturation of an OPC-concrete with w/c-ratio 0,54 and air content 7,1 %. E_n is the dynamic E-modulus after n freeze/thaw cycles; [16].

The number of freeze/thaw cycles seems to have a negligible effect on the S_{CR}-value, which means that there is almost no effect of fatigue. One example is shown in Fig. 4. The S_{CR}-value determined by the onset of expansion is well-defined already after 1 freeze/thaw cycle, [27]. This fact has also been demonstrated in other tests, [10], [28]. This negligible effect of repeated freeze/thaw cycles depends on the fact that the specimen is sealed during the test. In an unsealed test, there is normally increasing damage with increasing number of cycles; e.g. in the traditional salt scaling test. But, this is probably almost entirely depending on the fact, that moisture flows into the specimen during the test. This gradually increases its degree of saturation.

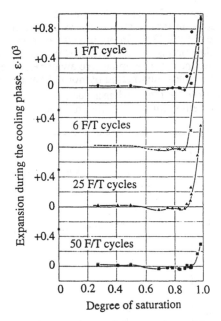

Fig. 4. Effect of the degree of saturation and the number of freeze/thaw cycles on the expansion of a concrete during freezing, [27].

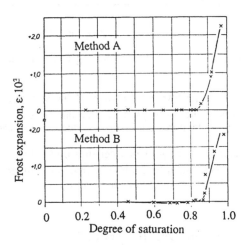

Fig 5. Effect of the degree of saturation and the freezing rate on the expansion of a concrete during freezing (Method A, 6°C/hour; Method B, 12°C/hour); [27].

The S_{CR}-value is also almost independent of the freezing rate. One example is seen in Fig. 5 showing the effect of the freezing rate on the expansion of specimens with different but constant degrees of saturation, [27]. Another example is shown in Fig. 6, which shows the result of a comparative international investigation of the so called S_{CR}-method; [21]. An analysis performed in [29] explains this behaviour theoretically.

In normal freeze-testing, an effect of the freezing rate is often found. The results are, however not unambiguous; in some tests an increased freezing rate brings about an increase of the damage; e.g. [30], [31]. In other tests, the opposite is the case; e.g. [32]. The most plausible explanation is that different test procedures affect in different ways the possibility of the specimen to take up or give away moisture; [29].

Thus, S_{CR} seems to be almost unaffected by normal variations in the environmental conditions. It can, therefore, be regarded as a true material property, a fact that is utilized in the the proposed method for service life prediction.

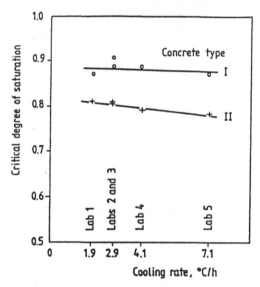

Fig. 6. Effect of the freezing rate on the critical degree of saturation of two OPC-concretes; [29].

It might be, that the S_{CR}-value is a function of the salt concentration of the pore water, and that a lowest value is obtained for a certain pessimum concentration; [33]. The effect of salt water in the pores on the critical degre of saturation, therefore, has to be clarified.

One can also expect, that the S_{CR}-value depends on the lowest temperature used in the test. Thus, it has been found that a lowest freezing temperature of -22°C causes considerably more scaling in a traditional salt scaling test than the temperatures -8°C and -14°C; [34].

7 The actual degree of saturation - the service life

In Fig. 7, a representative volume inside the concrete is shown. The S_{CR}-value can be regarded constant after the first few months, provided there is no change of the air-pore structure, such as deposition of crystals in the "active air-pores", pores that are coarse enough not to become readily water-filled during normal conditions. The actual degree of saturation, S_{ACT}, of the unit volume, and the temperature, changes with changes in the external climate. At point B, the unit volume freezes, at the same time as the S_{ACT}-value of the unit volume exceeds its S_{CR}-value. This causes considerable destruction of the unit volume. One can assume, that a large number of adjacent unit

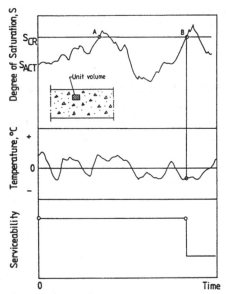

Fig. 7. Hypothetical time functions of the actual degree of saturation, S_{ACT}, the critical degree of saturation, S_{CR}, and the temperature inside a unit volume in a concrete member.

volumes are damaged at the same time giving a measurable reduction of the serviceability of the entire structure.

The frost resistance problem is a good example of the statistical nature of durability. The unlucky combination of excess of water and low temperature might occur during the first year, as well as after many decades. Therefore, it is very difficult -or even impossible- to make an exact prediction of the service life of concrete exposed to natural frost action. It is evident, however, that a long service life implies that the concrete is designed in such a way that the probability is very low that S_{ACT} will ever exceed S_{CR}.

8 The potential service life - the capillary absorption

Instead of trying to analyze, or simulate, the real environment, which is practically impossible, but if successful, should give the true service life, one can utilize a "standard environment", and obtain a sort of "potential service life". One possibility is to use a long-term water absorption test, and measure the capillary degree of saturation, S_{CAP} as function of the suction time. Thus, the true and fluctuating value S_{ACT} is replaced by the gradually increasing value S_{CAP}. The potential service life t_p is defined

$$S_{CAP}(t_p) = S_{CR} \qquad (7)$$

Where $S_{CAP}(t_p)$ is the degree of saturation after the time t_p of continuous capillary water uptake. In theory, the time function $S_{CAP}(t)$ could be calculated by means of advanced moisture mechanics, and a detailed knowledge of the air-pore system. Such a theory is worked out in detail in [20]. Examples of an application of the theory are shown in Fig. 8, where the degree of water-filling S_a of the air-pore system is plotted as function of time. $S_a=0$ and $S_a=1$ correspond to an empty and a completely water saturated air-pore system respectively. Thus, water in the capillary pores and gel pores are not included in S_a. The assumed frequency function of the air pores is.

$$f(r) = n \cdot [\ 1/r^b - 1/r_{max}{}^b] \tag{8}$$

Where n and b are constants, b being a function of the specific area α_0 of the entire air-pore system. r_{max} is the radius of the coarsest air pore. The theoretical solution (see Eqn (13) below) is performed for two different values of the diffusivity of dissolved air through pore water. With a diffusivity of 10^{-12} m^2/s, and a specific area of 30 mm^{-1}, the time needed to fill 50 % of the air-pore system is about 30 years. The finer the pore system, the more rapid the rate of water-filling, and the shorter the service life. With a specific area of 50 mm^{-1} it only takes 1 year to fill 50 % of the air-pore system. The theoretical calculations were compared with experimental tests yielding fairly good agreement.

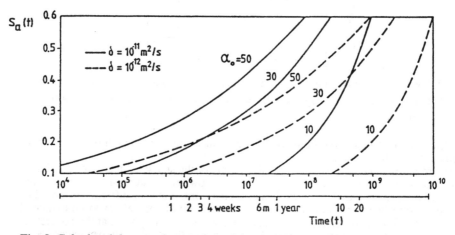

Fig. 8. Calculated degree of saturation of the air pore system of concretes permanently stored in water and with air pore systems that are described by Eqn (8); [20].

A practical experimental method for determination of S_{CAP} consists of a capillary absorption test using thin concrete plates (about 20 mm), that are completely immersed in water for a long time, or that are sucking water from one side, which is put in contact with the surface of water contained in a vessel, that is covered by a lid in order to protect the top surface of the specimen from evaporation. The thickness of the plate must be small so that the measured water absorption can be assumed to occur homogeneously over the entire material volume. Typical absorption-time curves are shown in Fig. 9. The nick-point in the diagram, at point $(t_n; S_n)$, corresponds to the stage where all capillary pores and gel pores are completely water-filled, while no air-pores are filled. Thus, the quantity $(1-S_n) \cdot \varepsilon$ is equal to the total air content a_0 of the concrete. The good agreement between these two quantities is seen in Fig. 10.

The water absorption, that occurs after the nick-point, corresponds to the gradual water-filling of the air-pore system. This is caused by dissolution in the pore water of air contained in the pores, and transfer of this air by diffusion to coarser air pores, and to the specimen surface. The water absorption process after the nick point absorption can be described by an equation of type (9).

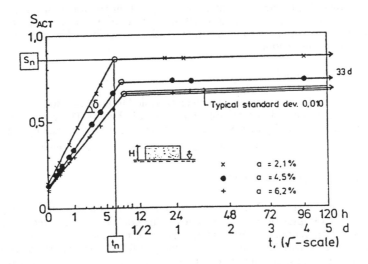

Fig. 9. Examples of the water absorption curves of OPC-concrete slices with one surface put in contact with water (a is the air content of the hardened concrete);
[18].

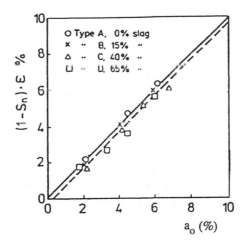

Fig. 10. Relation between the parameter $(1-S_n)\cdot\varepsilon$ and the total air content a_0 of concretes made with different types of cement and with w/c ratio 0,45; [18].

$$S_{CAP}(t) = [1/\varepsilon]\cdot\{\varepsilon_0 + f[a_0;w/c;cem.type]\cdot g[f(r);t]\} \qquad (9)$$

where ε_0 is the porosity exclusive of air pores. The total porosity is $\varepsilon=\varepsilon_0+a_0$. The
first term within the second paranthesis, $\varepsilon_0/\varepsilon\approx S_n\cdot\varepsilon$, therefore, is the rapid absorption in
the gel, and capillary pores up to the nick-point, and the second term determines the
absorption in the air-pores. The function f describes the rate of water absorption in the
air-pores, and the function g describes the shape of the time-curve of this process.

Our knowledge of the two functions f and g is limited, although some estimates can be made as seen in Fig. 8. One possibility is to extrapolate the result of a water absorption test which is run for some time (weeks or months). The most simple extrapolation is a lin-log relation.

$$S_{CAP}(t) = A + B \cdot \log t \qquad (10)$$

where A and B are constants, which are adjusted to the test results, and where t is the suction time expressed in a suitable unit (normally hours). The coefficient A is almost the same as the quantity $\varepsilon_o/\varepsilon \approx S_n \cdot \varepsilon$, and B is determined by the rate by which air in the air-pores can dissolve and be replaced by water. Eqn (10) can, therefore, be written.

$$S_{CAP}(t) = [1/\varepsilon] \cdot [\varepsilon_o + B \cdot \varepsilon \cdot \log t] \qquad (11)$$

where the coefficient B involves all the material parameters that are implicit in Eqn (9). Some examples of measured values of the coefficient B are shown in Fig. 11. B increases with increasing air content, which is reasonable, since the diffusivity of gases increases with increasing air content; [20]. B also increases with increasing amount of ground granulated blast furnace slag in the cement. The reason is not fully known. It was found, however, that the specific area of the air-pore system increased substantially with increasing slag content. Theoretically, this should give the observed effect on the coefficient B; see Fig. 8.

Combination of Eqns (7) and (10) gives the following relation for the potential service life.

$$t_p = 10^{(SCR-A)/B} \qquad (12)$$

A theoretical analysis indicates, that a lin-log extrapolation might overestimate the service life; viz. the real long-term absorption seems to be more rapid, than that predicted by Eqn (10); [20]. A relation of the following type seems to be more realistic for the degree of saturation S_a of the air-pore system itself, provided $S_a < 0,5$.

$$S_a(t) = C \cdot \alpha_o^D \cdot (\delta \cdot t)^E \qquad (13)$$

where C and D are general constants, E is a constant, which also depends on the specific area of the air pore system, α_o, and δ is the diffusivity of dissolved air.

The capillary degree of saturation of the total concrete is.

$$S_{CAP}(t) = [1/\varepsilon] \cdot [\varepsilon_o + S_a(t) \cdot a_o] \qquad (14)$$

Inserting Eqn (13) gives.

$$S_{CAP}(t) = [1/\varepsilon] \cdot [\varepsilon_o + C \cdot \alpha_o^D \cdot a_o \cdot (\delta t)^E] \qquad (15)$$

This can also be formulated

$$S_{CAP}(t) = A + F \cdot t^E \qquad (16)$$

where the constant $F = [C \cdot \alpha_o{}^D \cdot a_o \cdot \delta \ E]_{/\varepsilon}$ is individual for each concrete. A is the same as in Eqn (10). The potential service life then becomes

$$t_p = [(S_{CR} - A)/F]^{1/E} \tag{17}$$

Provided one has information on how different concrete technology parameters such as w/c-ratio, cement type, curing, moisture history, etc, affects the functions f and g in Eqn (9), and the critical distance L_{CR} (or D_{CR}), it should be possible to make a purely theoretical calculation of the potential service life of a concrete stored in water, and then frozen. Neither a freeze/thaw test, nor a time consuming water absorption test would be necessary. The only information needed concerns the air-pore structure. Then, one can calculate the S_{CR}-value according to the principles described in paragraph 5, and the S_{CAP}-function by Eqn (15). A combination yields the potential service life.

It is quite clear, that by using the S_{CR}-concept, one can, from a theoretical point of view, obtain quantitative information of the expected service life, and not only a rough qualitative durability level as that obtained by a traditional freeze/thaw test. One can also, by making parameter studies, investigate the sensitivity with regard to frost resistance of different material parameters, such as pore size distribution, air content, w/c ratio, diffusivity (permeability), etc.

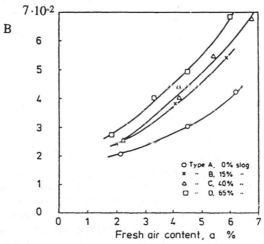

Fig. 11. Examples of experimental determinations of the coefficient B in Eqn (10) as function of the air content and cement type; [18] (w/c ratio 0.45)

9 The required air content for a given service life

The required total air content a_o is normally calculated by Powers equation (1) expressed in the following form, based on the specific area α_o of the empty (dry) air pore system, and the fictitious critical spacing factor $(L_o)_{CR}$.

$$a_o = V_p / \{0.364 \cdot [(L_o)_{CR} \cdot \alpha_o/3 + 1]^3 - 1\} \tag{18}$$

As said above, from a theoretical point of view, no value of $(L_o)_{CR}$ can exist. A certain fraction of the air-pore system will always become water-filled during natural conditions. The fact that certain researchers seem to have found rather well-defined values of $(L_o)_{CR}$ for normal types of concrete most likely depends on the fact that pore systems in such concretes often have similar shape. Therefore, possibly there exists a fairly constant relationship between the fictitious value $(L_o)_{CR}$ and the true value L_{CR}. This relation will probably no longer exist for air-pore systems of divergent appearance, such as extremely fine-porous systems, or extremely coarse-porous systems. In the first case, the air requirement calculated by means of Eqn (18) will be underestimated due to an underestimation of the substantial water absorption in such pore systems. In the second case, the air requirement will be overestimated because such pore systems do not absorb much water; almost all pores will stay air-filled even during very moist conditions.

A more realistic value of the required air content should be calculated by the following equation.

$$a_o = a_w + a_{CR} + a_b \qquad (19)$$

where a_w is the water-filled air pore volume, a_{CR} is the air pore volume needed in order that the spacing L_{CR} should not be exceeded when a_w is reached in the pore system, and a_b is an "air-buffer", or a safety margin. The value a_{CR} is calculated by the Powers equation (2) using the true value L_{CR} and the value α_{CR} which is the specific area of the air-filled part of the air-pore system when the residual spacing of this is exactly L_{CR}. Then, Eqn (2) can be written

$$a_{CR} = V_p/[0.364 \cdot (L_{CR} \cdot \alpha_{CR}/3 + 1)^3 - 1] \qquad (20)$$

The value a_w is depending on the environmental conditions. The wetter the environment, the higher the value of a_w and the bigger the size of the largest water-filled air-pore. Therefore, the value of the "residual" specific area α_{CR} is reduced when the environment becomes wetter, and according to Eqn (20), the required air volume a_{CR} is increased. Therefore, according to Eqn (19), in order to compensate for a wetter environment, the total air content ao of the concrete must be increased. This is not considered when the air requirement is calculated by Eqn (18), since neither the value of $(L_o)_{CR}$, nor the value of α_o are dependent of the wetness of the environment.

The ideas presented are illustrated by Fig. 12 showing the effect of a gradual water-filling of the air-pore system on the residual values L_r, a_r and α_r of the parameters L, a and α. A water-filling, that corresponds exactly to the residual spacing factor L_{CR} also gives the residual specific area α_{CR} and the residual air content a_{CR}. All pores with a diameter smaller than Φ_{CR} are then water-filled. In the real case, only pores smaller than Φ_w are water-filled. Let us assume that $\Phi_w < \Phi_{CR}$. This gives an air-buffer a_b, which is a sort of safety margin against the occurrence of exceptionally moist conditions. The air-buffer will increase with increasing air content under the assumption that the shape of the air-pore system is unchanged. A long service life requires an air-buffer, which is large enough never to be fully utilized even during very moist conditions.

The increased air requirement when freezing occurs in the presence of salt can possibly be understood by the theory just presented. The extra need of air is normally explained by the fact that salt apparently reduces the critical fictitious factor $(L_o)_{CR}$ from about 0.25 mm to about 0.16 to 0.20 mm for normal concrete. Inserted in Eqn (18) this decreased $(L_o)_{CR}$-value gives a substantial increase in the required air content.

The following reasoning is however just as plausible. The presence of salt increases the water content in the air-pore system of the surface part of the concrete, e.g. by attracting water from the interior of the concrete, or by prolonging the periods of wetness of the concrete surface. Thus, the value a_w increases, as well as the size Φ_w of the largest water-filled air-pore. This means that the specific area of the air-filled part of the pore system decreases. Then according to Eqn (20) the required air content a_{CR} of that portion of the air pore system, that has to stay air-filled, either increases, or decreases, depending on the effect of salt on the true spacing factor L_{CR}. The decrease in the specific area α_{CR} might be more or less compensated for by the possible increase in the true L_{CR}-value, that was found in the tests described in paragraph 3. The combined effect of the changes in a_w and a_{CR} is such, however, that the total air requirement increases.

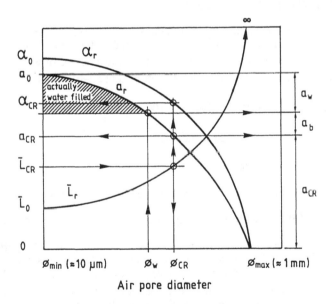

Fig. 12. Illustration of the principles behind the concepts residual specific area α_r, residual air content a_r, residual spacing factor L_r, water-filled air pore volume a_w, critical air pore volume a_{CR}, and air buffer a_b.

The application of the theory is shown by some examples. In all examples the following concrete is assumed: The cement paste fraction (including "interfering" fine aggregate) is 0.37 m^3/m^3, the air void system is described by Eqn (3) with the constant b=1.03 when the pore radius is expressed in μm. This corresponds to a specific area of the empty pore system of 30 mm^{-1}. The calculations of the relations between the waterabsorption a_w, the corresponding pore diameter Φ_w, and the residual specific area α_r are based on formulas presented in [15].

The true critical spacing factor is assumed to have the values, that were found experimentally (paragraph 3); 0.40 mm for freezing in water and 0.54 mm for freezing in salt.

Example 1: No salt. Short storage in water

All pores with diameter smaller than 140 μm are assumed to be water-filled. This corresponds to a residual specific area $\alpha=23$ mm^{-1} and a degree of water-filling S_a of the air pore system of 16 %. The required amount of air-filled pores is; Eqn (20).

$$a_{CR} = 0.37/[0.364 \cdot (0.40 \cdot 23/3+1)^3 -1] \cdot 100 = 1.6 \%$$

The required minimum total air content is; Eqn (19).

$$(a_o)_{min} = S_a \cdot (a_o)_{min} + a_{CR} = 0.16 \cdot (a_o)_{min} + 1.6 = 1.9 \%$$

Example 2: No salt. Long storage in water

The amount of water in the air pore system is higher than in example 1. All pores with diameter smaller than 200 μm are assumed to be water-filled. This corresponds to a residual specific area of 19.5 mm^{-1} and a degree of water-filling S_a of the air pore system of 34 %. The required amount of air-filled pores is.

$$a_{CR} = 0.37/[0.364 \cdot (0.40 \cdot 19.5/3+1)^3-1] \cdot 100 = 2.3 \%$$

The required minimum total air content is:

$$(a_o)_{min} = S_a \cdot (a_o)_{min} + a_{CR} = 0.34 \cdot (a_o)_{min} + 2.3 = 3.5 \%$$

Thus, the air content must be increased by 1.6 % when the concrete is more moist.

Example 3: With salt. The same long storage in water as in example 2

The water absorption is assumed to increase when salt is used. All pores with diameter smaller than 300 μm are assumed to be water-filled. This corresponds to a residual specific area of 15 mm^{-1} and a degree of water-filling S_a of the air pore system of 65 %. The true critical spacing factor is now assumed to be increased to 0.54 mm. The required amount of air-filled pores, therefore, is.

$$a_{CR} = 0.37/[0.364(0.54 \cdot 15/3+1)^3-1] \cdot 100 = 2.1 \%$$

This is almost the same value as for freezing in pure water; the decrease in α is compensated for by the increase in L_{CR}. The required minimum air content is:

$$(a_o)_{min} = S_a \cdot (a_o)_{min} + a_{CR} = 0.65 \cdot (a_o)_{min} + 2.1 = 6 \%.$$

Thus, the required air content might be higher when freezing occurs in the prese ce of salt despite the fact that the true critical spacing factor might be increased.

10 Suggestions for future research

In order to verify the theory, and make it more applicable also when salts are involved, the following experimental studies -among others- should be made.

* Determination of the true critical spacing factors from measurements of S_{CR}-values of concretes containing salt water of different concentrations, and with different air pore structures. The evaluation is made according to the principles described in paragraph 4.

* Determination of the long term water absorption in air-pore systems of different shape, and during different outer conditions; varying temperature, interrupted and resumed water absorption, variable salt concentration etc.

* Determination of the "micro-flow" of salt and water in the surface part of a concrete during a freeze/thaw cycle, and between the surface and the surrounding medium; pure water or salt water.

Much of the work deals with moisture flow, which is natural because the frost resistance problem is to a very large extent a "moisture mechanics problem". One will never obtain a good theory for the service life with regard to frost attack if the moisture mechanics of the problem is not understood.

12 References

1. Powers, T.C. (1949) The air requirement of frost resistant   concrete. Proc. Highway Research Board, 1949, 12, 184-211.
2. Powers, T.C. (1956) Resistance of concrete to frost at early ages. Proceedings RILEM Symposium on Winter Concreting. Copenhagen 1956, Session C, 1-46.
3. Powers, T.C. The mechanism of frost action in concrete. Stanton Walker Lecture Series on the Materials Sciences.Lecture No 3. Univ. of Maryland, 1965, 35 pp.
4. Powers, T.C., Helmuth, R.A. (1953) Theory of volume changes in portland cement paste during freezing. Proc. Highway Research Board, 1953, 32, 285-297.
5. Everett, D.H. (1961) The thermodynamics of frost damage to porous solids, Trans. Faraday Society, 1961, 57, 1541-1551.
6. Haynes, J. M. (1964) Frost action as a capillary effect. Transaction of the British Ceramic Society, 1964, 11, 697-703.
7. Dunn, J.R., Hudec, P.P. (1965) The influence of clay on the water and ice in rock pores. Physical Research Report RR 65-5, New York State Department of Public Works, 1965.
8. Setzer, M.J. (1977) Einfluss des Wassergehaltes auf die Eigenschaften des erhärteten Betons, Schriftenreihe DAStb. 1977, Heft 280.
9. Litvan, G.G. (1972) Phase transitions of adsorbates: IV Mechanisms of frost action on hardened cement paste. J. American Ceramic Society, 1972, 55, 38-42.
10. Fagerlund, G. (1986) The critical size in connection with freezing of porous materials. Cementa Report CMT 86039, Stockholm, 1986, 14pp. (In Swedish).
11. Ivey, D.L., Torrans, P.H. (1970) Air void systems in ready mixed concrete. J. of Materials, 1970, 5, 492-522.
12. Bonzel, J., Siebel, E. (1977) Neuere Untersuchungen über den Frost-Tausaltz-Widerstand von Beton, Beton,1977,Heft 4, 153-157,Heft 5 205-211,Heft 6 237-244.
13. Gagne, R., Aitcin, P.C., Pigeon, M., Pleau R. (1992) Frost durability of high performance concrete, in High Performance Concrete. From material to structure (Ed.Y. Malier), F&FN Spon, London, 1992, 239-251.

14. Löfgren, M., de Scharengrad, P. (1991) Salt-frost resistance of air-free cement paste. Report TVBM-5021, Div. of Building Materials, Lund Inst. of Technology, Lund, 1991, 72 pp. (In Swedish).
15. Fagerlund, G. (1979) Prediction of the service life of concrete exposed to frost action. In Studies on Concrete Technology, Swedish Cement and Concrete Research Institute, Stockholm, 1979, 249-276.
16. Fagerlund, G. (1981) The principles of frost resistance of concrete. Nordisk Betong, 1981, 2, 5-13. (In Swedish with English summary).
17. Fagerlund, G. (1978) Frost resistance of concrete with porous aggregate, Report Fo2:78, Swedish Cement and Concrete Research Institute, Stockholm, 1978, 189 pp.
18. Fagerlund, G. (1982) The influence of slag cement on the frost resistance of concrete. Report Fo1:82, Swedish Cement and Concrete Research Institute, Stockholm, 1982, 111 pp.
19. Warris B. (1964) The influence of air entrainment on the frost-resistance of concrete, Proceedings No 36, Swedish Cement and Concrete Research Institute, Stockholm, 1964, 130 pp.
20. Fagerlund, G. (1993) The long time water absorption in the air pore structure of concrete. Report TVBM-3051, Div. of Building Materials, Lund Institute of Technology, Lund, 1993, 75 pp.
21. Fagerlund, G. (1977) The critical degree of saturation method of assessing the freeze/thaw resistance of concrete and The international cooperative test of the critical degree of saturation method of assessing the freeze/thaw resistance of concrete, Materials and Structures, 1977, 10, 217-229 and 231-253.
22. Fagerlund, G. (1972) Critical degrees of saturation at freezing of porous and brittle materials, Report 34, Div. of Building Technology, Lund Institute of Technology, Lund, 1972, 408 pp.
23. Fagerlund, G. (1975) The significance of critical degrees of saturation at freezing of porous and brittle materials, in ACI Special Publication SP-47, Detroit, 1975, 13-65.
24. Studer, W. (1979) Versuche über die Anwendbarkeit der Frostbeständigkeitsprüfung nach "Tentative Recommendation RILEM Committee 4 CDC", EMPA Untersuchungsbericht No 10´780, Dübendorf, Switzerland, 1979.
25. Studer, W. (1980) Die Bestimmung der Frostbeständigkeit von Beton, Schweizer Bauwirtschaft, 1980, 20, 1-4.
26. Stehno, G. (1979) Practical experiments making use of the critical degree of saturation method to determine the freeze/thaw resistance of concrete, in Proc. RILEM conference Quality Control of Concrete Structures, Stockholm, 17-21 June 1979, Swedish Cement and Concrete Research Institute, Stockholm 1979, 283-292.
27. Klamrowski, G., Neustupny, P. (1984) Untersuchungen zur Prüfung von Beton auf Frostwiderstand, Bundesanstalt für Materialprüfung, Forschungsbericht 100, Berlin 1984, 84pp.
28. Fagerlund, G. (1980) Testing of frost resistance, in Int. Colloquium on the Frost Resistance of Concrete, June 1980, Mitteilungen der Österreichischen Zementfabrikanten, Heft 33, Vienna, 1980, 60-86.
29. Fagerlund, G. (1992a) Effect of the freezing rate on the frost resistance of concrete. In Nordic Concrete Research, Publ No 11, Oslo, February, 1992, 10-36.
30. Nischer, P. (1976) Der Einfluss der Abkühlgeschwindigkeit auf das Ergebnis der Prüfung von Beton auf Frost-Tausalz-Beständigkeit, Zement und Beton, 21, Heft 2, 1976, 73-77.
31. Pigeon, M., Prévost, J., Simard, J-M. (1985) Freeze-thaw durability versus freezing rate, J. of the American Concrete Institute, Sept.-Oct, 1985, 684-692.

32. HRB (Highway Research Board) (1959) Report on cooperative freezing-and-thawing tests of concrete, Highway Research Board, Special Report 47, Washington D.C, 1959, 67 pp.
33. Fagerlund, G. (1992) Studies of the scaling, the water uptake and the dilation of mortar specimens exposed to freezing and thawing in NaCl-solution. In Freeze-thaw and De-icing Resistance of Concrete. Research seminar held in Lund, June 1991, Report TVBM-3048, Div. of Building Materials, Lund Institute of Technology, Lund, 1992, 35-66.
34. Lindmark, S. (1993) Unpublished results, Div. of Building Materials, Lund Institute of Technology, 1993.

3

High strength concrete without air entrainment:effect of rapid temperature cycling above and below 0°-C.

E.J. SELLEVOLD and J.A. BAKKE
The Norwegian Inst. of Technology, Trondheim, Norway

S. JACOBSEN
The Norwegian Building Research Institute, Oslo, Norway

Abstract
The purpose of the work was to test the hypothesis that thermal incompatibility between aggregate and binder phase in high strength concrete leads to deterioration under rapid temperature cycles even in a temperature range where no ice formation takes place. Concrete with water/binder - ratio of 0.30 containing 8 % silica fume was produced with 3 aggregate types: Quartz Diorite, Limestone and lightweight aggregate (Macrolite). All three concretes deteriorated after ASTM C666 testing, while temperature cycling between + 5 °C and + 27 °C led to small damage (durability factor = 92) for Limestone, and no damage for the other two aggregates. It does not appear that thermal incompatibility between binder phase and aggregate is a major cause of the deterioration commonly observed for high strength concrete without airentrainment in ASTM C666 freeze/thaw testing.
Keywords: Durability, freeze/thaw testing, thermal incompatibility, ice formation.

1 Background and purpose

The question of whether high strength concrete (HSC) needs air entrainment for frost resistance is still unsettled. For given materials there probably is agreement that below a certain water-binder ratio no air entrainment is needed, however, where this limit is has by no means been established. The choice of test method has been found to be very important, for instance in our experience HSC is very robust against salt/frost scaling [1, 2, 5], while volume deterioration can be marked and rapid under exposure to ASTM C666 procedure A [1]. Calorimetric measurements of ice formation on equivalent HSC showed that very little ice forms in the temperature range of the ASTM C 666 test. For example, in HSC with water - binder ratio of 0.30 containing 10 % silica fume only

Freeze-Thaw Durability of Concrete. Edited by J. Marchand, M. Pigeon and M. Setzer.
Published in 1997 by E & FN Spon, 2–6 Boundary Row, London SE1 8HN, UK.
ISBN 0 419 20000 2.

about 4 % of the evaporable water froze in a specimen predried at 50 °C before being immersed in water under vacuum [1]. That so little ice should lead to volume deterioration was difficult to believe, and we therefore suggested that thermal incompatibility between the aggregate and the binder phase could create stresses large enough to cause deterioration during the rapid temperature variations in the ASTM C 666 test.

The present experiments were designed to test this hypothesis by subjecting HSC without air entrainment to standard ASTM C666 procedure A testing as well as a "frost-without-frost" test where the temperature range in the specimen was from + 5 °C to + 27 °C instead of the normal + 4 °C to - 18 °C. Procedure A was followed in both cases, i.e. the specimens were surrounded by water during the entire test.

The concretes were produced with the same water - binder ratio of 0.30 containing 8 % silica fume by weight of the cement. Each concrete had different aggregates: QD (Quartz Diorite), LS (Limestone) and LWA (Light Weight Aggregate, type Macrolite). The two first were normal weight aggregates chosen for their differences in thermal expansion coefficients. LS was expected to be lowest, QD highest, but their values were not determined. Macrolite is a glassy light weight aggregate with a particle density of about 0.80 kg/l. It was chosen to give minimum constraint to the binder thermal movements

2 Materials and concrete compositions

The cement was a high strength type (P30 - 4A, Norcem, Norway), the silica fume in slurry form (Elkem, Norway). A limestone filler was used in all three concretes. QD and LS was used in the whole range 0.125 - to 16 mm for the two normal weight concretes, while the LWA concrete contained LWA in the range 0.3 - 12.7 mm with normal sand in the 0.125 - 0.3 mm range. 9 - 12 kg/m^3 of superplasticizer (Mighty with a damper) was used to produce concrete with slump about 16 cm. All three concretes contained a binder volume of about 32 % (excluding air) while the fresh air content was about 2 % in the two normal weight concretes. The aircontent of the LWA-concrete could not readily be determined. The aircontents in the hardened concretes were determined by the PF-method [5] and found to be about 1.5 % for QD and LS; again the method is not relevant for LWA-concrete. The specimens were cast as 100 x 100 x 1200 mm prisms, later sawn to lengths of 350 mm for frost testing. 100 mm cubes were also made for compressive strength determination. The specimens were cured in water for 4 months after demoulding, then wrapped in heavy plastic foil until testing, i.e. 8, 12 and 14 months respectively for the three test series carried out. The compressive strengths of the cubes were measured at 16 months age to be: QD = 141.6 MPa, LS = 110,5 MPa and LWA 41,0 MPa.

3 Frost experiments

The prisms were placed in opentopped thinwalled metal boxes surrounded on all sides by about 3 mm water. Plastic foils were placed in the water on top of the specimens to prevent evaporation of water during freeze/thaw testing. The freezer produced 5 cycles per 24 hours within the limits set by ASTM C666. A dummy specimen with a thermocouple in the centre and in the water surrounding the specimen was used to ensure proper cycle control. The deterioration of the concretes was measured by changes in the dynamic E-modulus, i.e. by resonant frequency.

Series I: Normal C666 Procedure A test, interrupted after 205 cycles because of equipment problems. 3 parallel specimens per concrete.

Series II: "Frost-without-frost" series with a temperature range from + 5 °C to +27 °C. 415 cycles were completed. 3 parallel specimens per concrete.

Series III: Series II resulted in very little deterioration in the concretes. 2 specimens from each concrete in series II were therefore subjected to 300 normal ASTM C 666 procedure A cycles.

Ultrasonic pulse velocity was also measured during the tests. In addition, the weights of the specimens were monitored, and a new method of crack detection (FLR = Fluorescent Liquid Replacement) [3] on polished sections was applied before and after the tests. The results have been reported [4]. Finally. low temperature calorimetry [1] was applied to QD-concrete both before and after freeze/thaw testing, to see if the deterioration was a consequence of increased ice formation as cracks developed.

4 Results and discussion

Figures 1, 2 and 3 give the results for individual specimens and mean values in each of the three test series. In Table 1, 2 and 3 all data are given. QD concrete (Figure 1) reaches a durability factor (DF) of 41 (average) after 200 cycles, while LS-concrete obtains DF = 71 and for ML - concrete DF = 25. Thus, as seen earlier, none of these HSC's can be considered frost resistant according to the ASTM C666 procedure A. Equivalent QD-concrete tested for salt/frost scaling [5] showed minimal scaling, even when cured at a temperature of 60 °C, dried at 50 °C and resaturated before testing. Note the datapoints at 3 months in Figure 1. Because of equipment problems these concretes were left at - 5 °C for 3 months. They were then heated to room temperature and the resonance frequency determined. As is clear from the figure a recovery has taken place. The phenomenon is presently being investigated further [6]. The weight data showed a clear relation between water uptake and frost deterioration. ML gained most weight, followed by QD and LS.

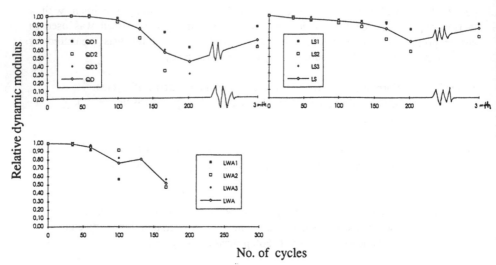

No. of cycles

Fig. 1. Series I (-18 °C to + 4 °C) ; Relative Dynamic Modulus vs. no. of cycles.

Figure 2 shows the results of Series II. After 415 cycles DF is 100 for QD - concrete, 92 for LS-concrete and 99 for ML - concrete. The scatter is very small for the two normal weight concretes, thus the deterioration in the LS - concrete appears significant, and is possibly the consequence of thermal incompatibility between binder phase and aggregate. However, seen together with the results of Series I it does not appear that this effect is very important in connection with freeze/thaw deterioration. Of course, the thermal expansion coefficient of these high strength binders over the entire range from - 20 °C to + 27 °C are not known.

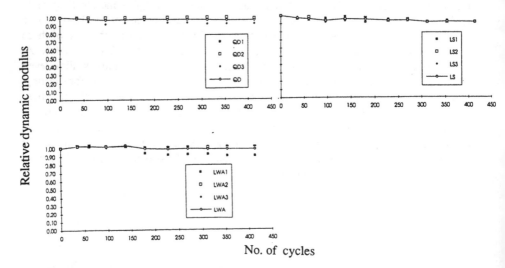

No. of cycles

Fig. 2. Series II (+ 5 °C to + 27 °C) ; Relative Dynamic Modulus vs. no. of cycles.

Figure 3 shows the results for the subsequent Series III. Mean DF was 79 for QD, 50 for LS and 23 for ML-concrete. These results are somewhat inconsistent with Series I for QD and LS concretes in that QD shows less deterioration in spite of the preceding "frost-without-frost" treatment., while LS is undisturbed up to about 170 cycles after which the deterioration is rapid. LS-concrete fails in both series, while QD - concrete fails in Series I, but does much better in Series II. We have no explanation for this contradiction. (A possible cause for the reduced durability of the LS - concrete in Series III could be cracks introduced during Series II, but again the improved durability of the QD-concrete is not understood).

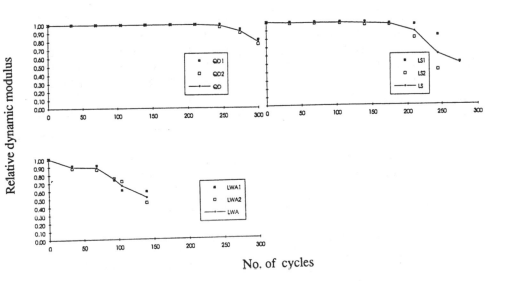

Fig. 3. Series III (-18 °C to + 4 °C) ; Relative Dynamic Modulus vs. no. of cycles.

The QD - concrete was tested in the calorimeter before and after freeze/thaw - testing. Small samples (d = 14 mm, l = 65 mm) were cut from the prisms and run in the existing moisture state. The total evaporable water contents were 0.0434 g/g_{wet} and 0,0414 g/g_{wet}, respectively, for the before and after samples. On cooling both showed a small amount of ice formation around - 3 °C, corresponding to 3.3 % and 3.5 % of the evaporable watercontent, respectively. Clearly the freeze/thaw treatment has not resulted in any major change in the ice formation during cooling in the temperature range covered by the ASTM C 666 freeze/thaw test. It is interesting to note, however, that the total amount of ice formed to - 55 °C increased from 10.8 % to 16.4 % of the evaporable water because of the freeze/thaw treatment.

Later work has shown that for higher w/c concrete that is more deteriorated in the ASTM C666 test ice formation increases over the whole temperature range [7].

Table 1. Series I (-18 °C to + 4 °C); Resonance Frequency and Mass.

Cycl	LWA1	LWA2	LWA3	LWA	LS1	LS2	LS3	LS	QD1	QD2	QD3	QD
						Freq.	(Hz)					
0	2450	2425	2450	2442	2875	2900	2950	2908	3325	2200	3300	3308
35	2425	2425	2450	2433	2825	2875	2900	2867	3325	3300	3300	3308
60	2350	2375	2425	2383	2800	2825	2925	2850	3300	3300	3275	3292
100	1850	2325	2225	2133	2800	2775	2875	2817	3275	3175	3225	3225
132			2200	2200	2775	2700	2850	2775	3225	2825	3050	3033
167		1675	1860	1763	2750	2450	2800	2667	2975	1900	2525	2467
202					2625	2175		2400	2600		1775	2188
300												
						Mass	(g)					
0	4994	4950	4949	4964	8710	8743	8675	8709	8830	8760	8845	8812
35	5022	4975	4972	4990	8710	8745	8677	8711	8838	8765	8852	8818
60	5063	5002	4993	5019	8711	8748	8678	8712	8835	8772	8857	8821
100	5160	5063	5061	5095	8716	8759	8683	8719	8851	8788	8871	8837
132	5302	5170	5123	5198	8723	8769	8686	8726	8858	8802	8884	8848
167		5234	5173	5204	8731	8784	8692	8736	8870	8821	8901	8864
202					8741	8798	8697	8745	8880	8835	8922	8879
300												

Table 2. Series II (+ 5 °C to + 27 °C); Resonance Frequency and Mass.

Cycl	LWA1	LWA2	LWA3	LWA	LS1	LS2	LS3	LS	QD1	QD2	QD3	QD
						Freq.	(Hz)					
0	2400	2400	2375	2392	2975	3000	3000	2992	3325	3300	3325	3317
36	2425	2425	2400	2417	2925	2950	2975	2950	3300	3300	3325	3308
60	2425	2425	2400	2417	2900	2975	2925	2933	3300	3300	3325	3308
96	2425	2425	2375	2408	2925	2900	2875	2900	3300	3300	3300	3300
138	2425	2425	2425	2425	2950	2925	2900	2925	3300	3300	3325	3308
180	2325	2425	2400	2383	2850	2950	2950	2917	3275	3300	3325	3300
228	2300	2425	2400	2375	2900	2900	2900	2900	3300	3300	3325	3308
270	2300	2425	2400	2375	2875	2925	2900	2900	3300	3300	3325	3308
312	2300	2400	2400	2367	2850	2875	2875	2867	3300	3300	3325	3308
354	354	2425	2400	2367	2875	2875	2875	2875	3275	3300	3300	3292
414	414	2425	2400	2367	2850	2875	2875	2867	3300	3275	3300	3292
						Mass	(g)					
0	5014	4988	4963	4988	8762	8855	8898	8838	8978	8945	8976	8966
36	5024	4993	4970	4996	8776	8861	8905	8847	8981	8947	8979	8969
60	5025	4995	4970	4997	8779	8863	8905	8849	8981	8948	8980	8970
96	5028	4994	4971	4998	8778	8862	8904	8848	8982	8948	8980	8970
138	5031	4997	4974	5001	8780	8864	8907	8850	8983	8950	8983	8972
180	5031	4998	4975	5001	8781	8866	8908	8852	8984	8952	8983	8973
228	5029	4994	4973	4999	8779	8865	8906	8850	8982	8950	8981	8971
270	5029	4996	4974	5000	8779	8866	8907	8851	8983	8951	8981	8972
312	5028	4995	4972	4998	8779	8866	8907	8851	8983	8950	8982	8972
354	5033	4999	4975	5002	8783	8869	8911	8854	8987	8953	8986	8975
414	5033	4998	4975	5002	8785	8870	8910	8855	8987	8954	8988	8976

Table 3. Series III (-18 °C to + 4 °C); Resonance Frequency and Mass.

Cycl	LWA1	LWA2	LWA	LS1	LS2	LS	QD1	QD2	QD
					Freq.	(Hz)			
0	2250	2425	2338	2875	2925	2900	3300	3300	3300
34	2150	2275	2213	2875	2900	2888	3300	3300	3300
69	2150	2250	2200	2875	2900	2888	3300	3300	3300
94	1925	2100	2013						
105	1750	2050	1900	2875	2900	2888	3300	3300	3300
140	1725	1625	1675	2875	2875	2875	3300	3300	3300
175				2850	2875	2863	3300	3300	3300
210				2825	2625	2725	3300	3300	3300
245				2625	1900	2263	3300	3250	3275
275				2050		2050	3200	3125	3163
300							2975	2875	2925
					Mass	(g)			
0	5032	4999	5016	8783	8868	8826	8985	8953	8969
34	5045	5010	5028	8784	8869	8827	8986	8953	8970
69	5077	5036	5057	8785	8870	8828	8986	8953	8970
94	5131	5065	5098						
105	5161	5086	5124	8785	8872	8829	8986	8953	8970
140	5174	5139	5157	8787	8875	8831	8985	8949	8967
175				8789	8881	8835	8987	8951	8969
210				8795	8883	8839	8988	8953	8971
245				8804	8909	8857	8992	8956	8974
275				8816		8816	8998	8963	8981
300							9003	8970	8987

5 Conclusions

High Strength Concrete (up to 140 MPa) without air entrainment performs badly in freeze/thaw testing according to ASTM C666 Procedure A, but is very resistant to salt/frost scaling according to Swedish Standard SS 13 72 44.

Rapid temperature cycles in the temperature range + 5 °C to + 27 °C causes some deterioration in concrete with Limestone aggregate (DF = 92), but none in concrete with Quartz Diorite.

Thermal incompatibility between aggregate and binder phase is not a major cause of disruption of high strength concrete during ASTM C666 testing.

The light weight aggregate concrete with the same high strength binder is particularly sensitive to ASTM C666 testing, probably because of water absorbed in the LWA.

Low temperature calorimetry did not show any increase in ice formation down to - 20 °C because of freeze/thaw testing to 300 cycles. The amount of freezeable water is only 3 - 4 % of the total evaporable water.

6 References

1. Hammer T.A. and Sellevold E.J.(1990) ACI SP-121 pp. 457 - 489.
2. Sellevold E.J and Farstad T.(1991) Nordic Concrete Research Publication No. 10 pp. 121 - 138
3. Gran, H.C. (1995) Cem.Conc.Res Vol.25 No.5 pp.1063-1074
4. Jacobsen S., Gran H.C., Sellevold E.J. and Bakke J.A.(1995) Cem.Conc.Res Vol.25, No.8, pp.1775-1780
5. Jacobsen S., Sellevold E.J.: "Frost/salt Scaling and Ice Formation of Concrete: Effect of Curing Temperature and Silica Fume on Normal and High Strength Concrete", paper part of this book
6. Jacobsen S. and Sellevold E.J.(1996) Cem.Conc.Res Vol. 26 No.1 pp. 55 - 62
7. Jacobsen S., Sellevold E.J., Matala S.(1996) Cem.Conc.Res. V.26 N.6 pp.913-931

Acknowledgement

The work reported here was part of the Norwegian research programme"High Strength Concrete, Materials Development". The program was financed by NTNF and industrial partners, with Norcem as project leader.

4

Scaling, absorption and dilation of cement mortars exposed to salt/frost

G. FAGERLUND
Lund Institute of Technology, Lund, Sweden

Abstract
Frost damage of concrete is normally much more severe when freezing takes place in the presence of dilute salt solutions. In this case, damage is almost always of type surface spalling or scaling, while the interior of the concrete remains unharmed. This effect of salts has been known for a very long time, for example in connection with the use of deicing salts or exposure to sea water. It is also known that more concentrated salt solutions cause less frost damage than do weak solutions. The mechanism behind this phenomenon is not very well understood. This study was, therefore, started in order to give some experimental input to the formulation of a destruction theory. Three types of test were performed with NaCl-solutions of four different strengths (0%, 2.5%, 5%, 10%); (i) a salt scaling test in which scaling, as well as water uptake, was monitored; (ii) an absorption test at constant room temperature; (iii) a dilation test during one-cycle freezing. Nine different cement mortars were tested; 3 water/cement ratios (0.4, 0.6, 0.7) and three air contents (natural air, and two different contents of entrained air). The main results are presented together with some comments.
Keywords: Frost resistance, salt scaling, water absorption.

1 Type of tests

Three types of test were made on different mature cement mortars:

Test 1: Surface scaling and salt water absorption of specimens that were freze/thaw tested when completely immersed in NaCl-solutions of four different concentrations; 0%, 2.5%, 5%, 10%. Nine different mortars were tested.

Test 2: Isothermal salt water uptake of specimens stored completely immersed in NaCl-solutions of the same concentrations as above. Nine different mortars were tested (the same as above.)

Test 3: Length changes during freezing of specimens that had been pre-conditioned to different degrees of saturation in NaCl-solutions of the same concentrations as above. Two different mortars were tested.

Freeze-Thaw Durability of Concrete. Edited by J. Marchand, M. Pigeon and M. Setzer.
Published in 1997 by E & FN Spon, 2–6 Boundary Row, London SE1 8HN, UK.
ISBN 0 419 20000 2.

2 Mortars

Nine mortars with the nominal water/cement ratios 0.4, 0.6 and 0.7 were made for Tests 1 and 2. The nominal air contents aimed at were 3%, 6%, and 8%. The 3% mortars were non-airentrained. Two different cement mortars with one water/cement ratio (0.6), and two air contents (3% and 8%) were made for test 3. The variables and the denomination of the different mortars are shown in Table 1.

The real water/cement ratios and air contens of the fresh mortars are shown in Table 2. The air contents differ quite much from the nominal, especially for the 6% mortars. Specimens for Test 1 and 2 were cast from one batch and specimens for Test 3 from another batch. Therefore, the recipes are a bit different.

The cement was of type low alkali, high sulfate resistant portland. The airentraining admixture was of the brand Darex AEA.

All specimens were table vibrated. The specimen size and the storage of the specimens were different for the two types of test; see below.

Table 1: Cement mortars. Denomination and basic variables (nominal)

Nominal air content	Water/Cement ratio		
%	0.4	0.6	0.7
3	11	12[1]	13
6	21	22	23
8	31	32[1]	33

1) Test 1, 2, and 3. All others, only Tests 1 and 2.

Table 2: Cement mortars. Mix proportions, w/c-ratios, air contents.

Mortar	Sand (kg/m^3)		Cement (kg/m^3)		w/c		Air (%)	
	Tests 1, 2	Test 3	Tests 1, 2	Test 3	Tests 1, 2	Test 3	Tests 1, 2	Test 3
11	1367		623		0.424		3.5±0.3	
21	1333		597		0.423		6.6±0.1	
31	1335		575		0.424		7.9±0.5	
12	1517	1530	383	412	0.642	0.600	7.1±0.3	3.7
22	1460		362		0.640		11.5	
32	1398	1550	347	378	0.605	0.600	15.0	8.4
13	1563		344		0.701		5.7±0.1	
23	1502		320		0.734		10.4	
33	1548		296		0.741		11.0	

3 Test 1. Salt scaling and salt water uptake

3.1 Aim
The aim of Test 1 was to investigate the effect of the outer salt concentration on the salt scaling and the absorption during freeze/thaw in NaCl-solution.

3.2 Test procedure
The specimens were pre-conditioned by two different methods:

Method 1: Water curing during 6 weeks, followed by air-drying for 2 weeks in laboratory atmosphere.
Method 2: Water curing for 6 weeks followed by air-drying during 1 week, followed by water storage during 2 weeks.

The pre-conditioned specimens (cylinders with diameter 50 mm, and length 100 mm) were placed in plastic beakers (diameter 82 mm, length 145 mm). About 0.53 litres of the salt solution (0%, 2.5%, 5%, or 10%) was poured into the beaker so that the specimen was completely covered by solution. The beaker was provided with a lid in order to prevent evaporation and thereby changes in the salt concentration during the test. Each specimen was exposed to 12 freeze/thaw cycles of the type seen in Fig 1.

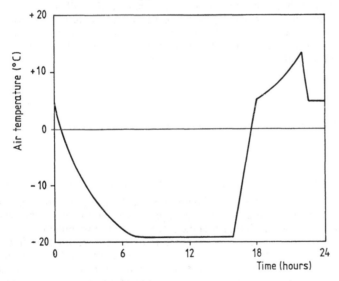

Fig. 1. Test 1. The freeze/thaw cycle

After terminated freeze/thaw test, the fragments, that were scaled off from the specimens, were collected in a paper filter, dried and weighed (ΔQ_d). The specimen was weighed before the test (Q_o) and after terminated freeze/thaw (Q_1). Then, it was dried at +105°C and weighed (Q_d). It was re-saturated by vacuum treatment, and weighed in water (Q_w) and in air (Q_a). Therefore, the porosity of the mortar (P) and the degree of saturation before freeze/thaw (S_o), and after freeze/thaw (S_1) could be calculated by the following equations:

$$P = \frac{Q_a - Q_d}{Q_a - Q_w} \tag{1}$$

$$S_0 = \frac{Q_0 - (Q_d + \Delta Q_d)}{P[Q_a - Q_w + \Delta Q_d / \gamma]} \tag{2}$$

$$S_1 = \frac{Q_1 - Q_d}{Q_a - Q_d} \tag{3}$$

Where γ is the bulk density of the cement mortar. The difference between S_0 and S_1 is a measure of the water absorption, or desorption, during the freeze/thaw test.

The S_0-values are almost the same for all specimens of the same type pre-treated in the same manner. One specimen of each mortar type was tested for each pre-conditioning type and each salt concentration. It is assumed that all weights are in grammes, and that no salt has been deposited in the pores. This is a reasonable assumption since the salt water storage during freeze/thaw is very short (12 days). In the calculations, no consideration is taken to the different densities of the different salt solutions. The error is rather small. The following densities were measured by aerometer for the different solutions:

$$\gamma_s = 0.9978 + 0.00658 \cdot c \tag{4}$$

Where γ_s is the density of the solution in kg/litre, and c is the concentration in %.

3.3 Results

The scaling ΔQ_d and the degree of saturation after freeze/thaw S_1 as function of the outer salt concentration in the beaker, are shown in Fig 2-4. In Fig 5-6 the scaling is plotted versus the water/cement ratio. Scaling is expressed in grammes for the whole specimen. In reality, scaling is much bigger at the lower pat of the specimen.

Freeze/thaw in salt solution always produce much more severe scaling than freezing in pure water. Besides, the NaCl-concentration 2.5% normally produces the most severe scaling of pre-dried specimens (pre-treatment according to Method 1). For this type of pre-conditioning, 10% solution is not much more harmful than freezing in pure water; see Fig 5.

When the specimens are water stored until freeze/thaw starts (pre-conditioning according to Method 2), 5% and 10% seem to the the most severe concentrations; Fig 6. One reason might be, that salt in this case has to migrate into the pores by a slow diffusion process, while it enters by a combined diffusion/convection process when the mortars were pre-dried. Hence, it migh very well be that the inner salt concentration in the mortar at its surface is the same for the two types of pre-conditioning. If this is the case, the most dangerous concentration *inside* the mortars will be the same for both types of pre-conditioning; i.e. about 2.5%.

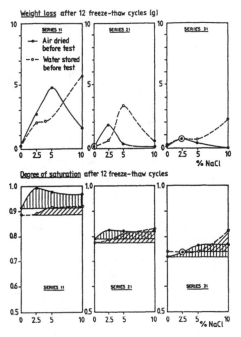

Fig. 2. Test 1. Mortars with water/cement ratio 0.4. Salt scaling and degree of saturation after 12 freeze/thaw cycles.

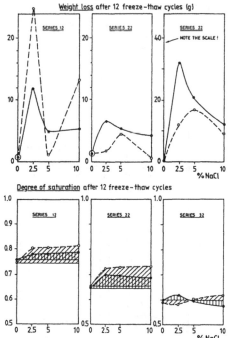

Fig. 3. Test 1. Mortars with water/cement ratio 0.6. Salt scaling and degree of saturation after 12 freeze/thaw cycles.

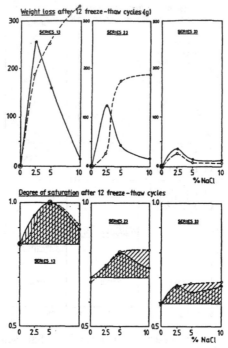

Fig. 4. Test 1. Mortars with water/cement ratio 0.7. Salt scaling and degree of saturation after 12 freeze/thaw cycles.

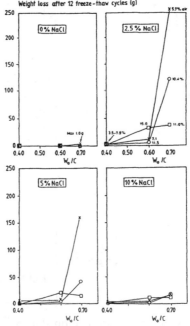

Fig. 5. Test 1. Pre-conditioning according to Method 1. Total weight loss after 12 freeze/thaw cycle

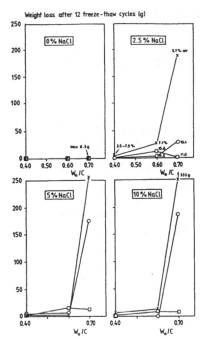

Fig. 6. Test 1. Pre-conditioning according to Method 2. Total weight loss after 12 freeze/thaw cycles.

The salt water absorption during the freeze/thaw cycles is normally lower when the specimens are frozen in pure water. One reason might be that the total time for water absorption is lower in this case because the specimen is surrounded by completely frozen water for more than half the test time, while it is constantly surrounded by a salt brine when the specimen is frozen in salt solution.

An additional explanation might be that the permeability and pore space of the remaining "torso" is increased when freeze/thaw takes place in salt water due to more severe frost damage. If this holds true, the observed increase in water uptake is just a measure of the increased internal damage occurring in salt water. It must be observed, however, that the degree of saturation is measured on the torso that is rather intact, and not on the outer part, that has scaled off. A high degree of saturation of the "torso" is, however, an indication of a high moisture content also in the destroyed part of the specimen.

The salt scaling is reduced when the air content is increased. When the air content is very high (e.g. 15% in mortar 32), there is a clear tendency of increased scaling. This is probably an effect of an air-pore system that was unstable in the fresh mortar. Thus, the air-pores might have formed a more or less continuous air-pore system, that was able to take up water during freeze/thaw. The air-pore structure was not investigated.

The scaling is also increased when the water/cement ratio is increased. When the water/cement ratio is 0.7, a very high air content is required for a frost resistant mortar. When the water/cement ratio is 0.4, also a non-airentrained mortar behaved fairly well in the test. Even in this case the scaling was, however, reduced when the mortars were airentrained.

The scaling was most severe at the lower part of the specimens. This phenomenon has not been explained. It is quite clear, however, that due to differences in the density between ice and solution, there will be a certain gradual stratifying within the solution as the cycles proceed. This will probably have an effect on the scaling.

4 Test 2. Isothermal salt water uptake

4.1 Aim
The aim of the test was to investigate the effect of the NaCl-concentration on the salt water absorption of air-dried specimens. 0%, 2.5%, 5%, and 10% NaCl-solutions were used.

4.2 Test procedure
The specimens were pre-conditioned according to Test 1, Method 1; i.e. they were water cured during 6 weeks followed by air-drying in laboratory air during 2 weeks.

The pre-conditioned specimens (diameter 50 mm, length 100 mm) were weighed (Q_o) and were then placed in the plastic beakers described for Test 1. Pure water or salt solutions was poured into the beakers so that the specimens were completely immersed. The total amount of solution was about 0.5 litres. The beakers were stored at room temperature for 13 days whereupon the specimens were weighed (Q_1). Then, they were dried at +105°C and weighed (Q_d). They were re-saturated by vacuum treatment and weighed in air (Q_a) and in water (Q_w). Two specimens were used for each concentration.

The degrees of saturation before (S_o) and after the test (S_1) are calculated by:

$$S_o = \frac{Q_o\text{-}Q_d}{Q_a\text{-}Q_d} \tag{5}$$

$$S_1 = \frac{Q_1\text{-}Q_d}{Q_a\text{-}Q_d} \tag{6}$$

Where Q is in grammes.

The density of the salt solution in the beaker before and after the test was determined by an aerometer and compared with the theoretical value according to eq (4). The solution that was sucked into the specimens was, however, so small in comparison with the total amount of solution surrounding the specimen, that the density measurements were too insensitive in order to reveal any concentration changes caused by an eventual slower transport of solved ions than transport of water.

4.3 Results
The degree of saturation after the terminated absorption test is plotted in Fig 7-9. For the water/cement ratio 0.7 it seems as if a higher salt concentration gives somewhat higher degrees of saturation when the air content is low, but lower degrees of saturation when the air content is high. The same trend, although less pronounced, can also be found for the other water/cement ratios. This means that it should be more difficult to water-fill entrained air-pores when salt solutions are used. There is no immediate explanation to this phenomenon.

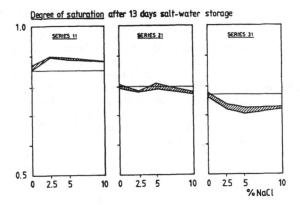

Fig 7: Test 2. Mortars with the water/cement ratio 0.4. The degree of saturation after 13 days of absorption at constant room temperature.

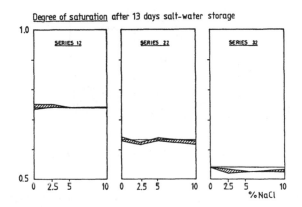

Fig 8: Test 2. Mortars with the water/cement ratio 0.6. The degree of saturation after 13 days of absorption at constant room temperature.

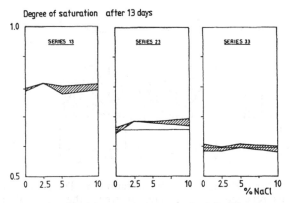

Fig 9: Test 2. Mortars with the water/cement ratio 0.7. The degree of saturation after 13 days of absorption at constant room temperature.

In Fig 10-12 is plotted the difference between the degree of saturation obtained during the freeze/thaw test (Test 1) and the degree of saturation obtained during isothermal water storage (Test 2). In almost all cases, freeze/thaw causes a bigger absorption than does isothermal storage. Besides, this "extra" salt water absorption caused by freeze/thaw is bigger when it takes place in salt water. One explantion is that frost damage during the very first cycles causes an increased porosity, that is readily filled during the additional freeze/thaw cycles. If frost damage did not occur, one might just as well have expected a lower water absorption during freeze/thaw due to the fact that the vapour pressure of a salt solution is lower than the vapour pressure of pure water. Thus, water ought to leave the specimen. In the freeze/thaw test, the concentration of the surrounding solution is on average much higher than in the isothermal storage. Thus, this drying effect should heve been bigger during freeze/thaw than during isothermal storage. Such a drying has been observed on mortar specimens stored in concentrated $CaCl_2$-solutions; [1], [2].

Possibly, some of the extra water absorption during the freeze/thaw test, could be explained by the cyclic temperature variation. Such an increased absorption in salt water has been observed for concrete stored in salt water under varying temperature above 0°C; [2].

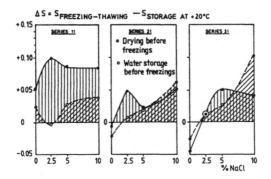

Fig. 10. Test 2. Mortars with the water/cement ratio 0.4. The difference in degree of saturations reached after 12 days of freeze/thaw and after 13 days of absorption at constant room temperature.

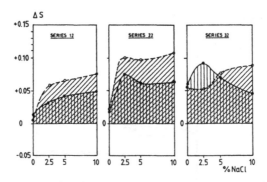

Fig. 11. Test 2. Mortars with the water/cement ratio 0.6. The difference in degree of saturations reached after 12 days of freeze/thaw and after 13 days of absorption at constant room temperature.

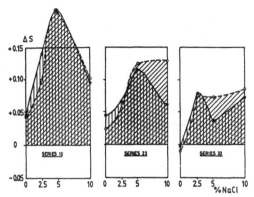

Fig. 12. Test 2. Mortars with the water/cement ratio 0.7. The difference in degree of saturations reached after 12 days of freeze/thaw and after 13 days of absorption at constant room temperature.

5 Test 3. Length changes during freezing

5.1 Aim
The aim of Test 3 was to investigate the effect of the inner salt concentration, and the effect of the degree of salt water saturation, on the expansion of the specimen during a single freezing.

5.2 Specimens and pre-treatment
All the 9 mortars used for Tests 1 and 2 were tested, but in this report only two mortars are treated; both with water/cement ratio 0.6, one without airentrainment (mortar 12) and one airentrained (mortar 32). See Table 2. The specimens were prisms with the cross-section 30·30·120 mm.

The specimens were water cured during 3 weeks followed by air-drying during 3 weeks. Then, each specimen was put in a vacuum chamber which was evacuated to a certain residual pressure, which was maintained for 24 hours. NaCl-solution of a given concentration (0%, 2.5%, 5%, or 10%) was let into the chamber while the pump was still running. The pump was turned off when the solution covered the specimen. Then, the specimen was stored in salt solution of the same concentration during 3 months.

This long storage, in combination with the forced absorption due to vacuum, made the pore water also in the centre of the specimen come into equilibrium with the outer concentration. This was confirmed by observations of the initial freezing temperatures obtained immediately after ice formation was initiated (This occurred after a certain super-cooling). These temperatures correspond very well to the freezing points that are expected in bulk solutions of the same concentration; See Fig 13.

By using different residual pressures during the evacuation phase, different degrees of salt water saturation (S) were reached. The following residual pressures were used:

1. 760 torr (no vacuuum): Degree of saturation S_1
2. 150 torr: S_2
3. 50 torr: S_3
4. 2 torr: S_4

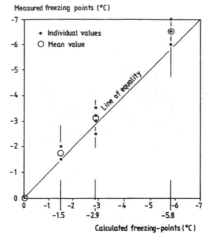

Fig. 13. Test 3. Measured initial freezing temperatures of the salt water stored specimens and the theoretical freezing points of the bulk solution.

5 3 Freeze test procedure

The pre-conditioned specimen was placed in the combined calorimeter/extensometer equipment shown in Fig 14; [3]. The simultaneous ice formation and length change was measured. Only length changes are shown in this paper. The calorimeter/extensometer was immersed in kerosene the temperature of which was gradually lowered to -28°C. The freeze curve is shown in Fig 15. The temperature was measured by a thermocouple at the centre of the specimen. Only one freeze/thaw cycle was used. The specimen was weighed before the test (Q_o) and after the test (Q_1). A very small amount of water was lost during the test; i.e. $Q_o \approx Q_1$. After the test, the specimen was dried at +105°C and weighed (Q_d). It was re-saturated by vacuum, and weighed in air (Q_a) and in water (Q_w). The degree of saturation during the test was calculated by eq (3) and eq (5).

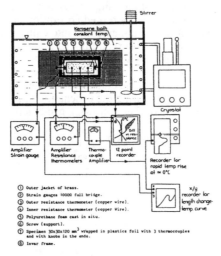

Fig. 14. Test 3. Combined calorimeter/extensometer.

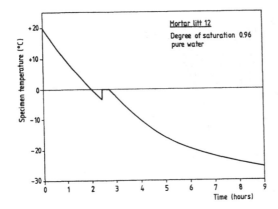

Fig. 15. Test 3. The freeze/thaw cycle.

5.4 Test results
The results are shown in Fig 16-19. In Fig 20 the total dilation at -25°C is plotted versus the salt concentration for differen constant degrees of saturation.
  The following conclusions can be drawn:

1: When the degree of saturation is low, no or very small expansion occurs for all salt concentrations.

2: When the degree of saturation is high, the largest expansion occurs at the concentration 2.5%. The concentrations 5% and 10% are not more harmful than pure water.

  Unfortunately, specimens with medium high degrees of saturation were not tested. Therefore, it is not possible to see the eventual effect of the salt concentration on the critical degree of saturation.

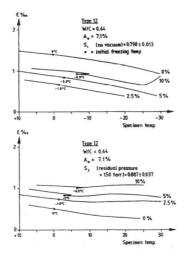

Fig. 16. Test 3. Length-temperature curves of mortar 12 (water/cement ratio 0.6, air content 3.7%). Low degrees of salt water saturation.

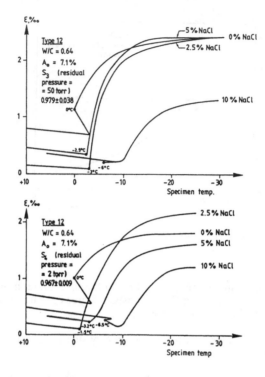

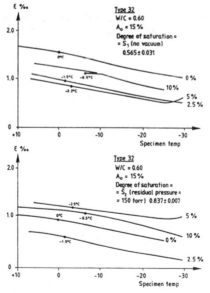

Fig. 17. Test 3. Length-temperature curves of mortar 12 (water/cement ratio 0.6, air content 3.7%). High degrees of salt water saturation.

Fig. 18. Test 3. Length-temperature curves of mortar 32 (water/cement ratio 0.6, air content 8.4%). Low degrees of salt water saturation.

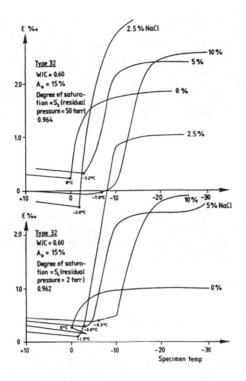

Fig. 19. Test 3. Length-temperature curves of mortar 32 (water/cement ratio 0.6, air content 8.4%). High degrees of salt water saturation.

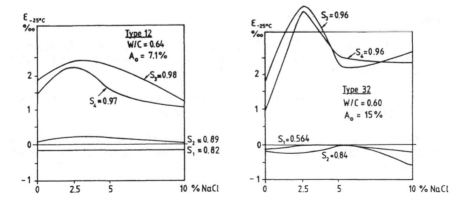

Fig. 20. The dilation at -25°C as function of the degree of salt water saturation and salt concentration.

Thus, it seems as if a 2.5% solution inside the specimen creates bigger destructive forces than the other concentrations used. One imaginable explanation is that the salt creates an osmotic pressure between water in pores containing ice bodies and pores with only unfrozen salt solution. This does not happen when the pore water is "pure". On the other hand, when the salt concentration is high, the freezable water decreases. Therefore, the number of sites where osmotic pressure appears is reduced. The net effect of different salt concentrations might be that the maximum stresses appear at a concentration that is moderately high. The worst case might be a concentration of the order 2 to 3%.

6. Conclusions

The results of the tests can be summarized as followes:

1: The most severe salt scaling of specimens, that are freeze/thaw tested in NaCl-solution, occurs at a certain pessimal concentration of the outer salt concentration. This is of the order 2.5% for specimens that are air-dried when freezing starts and 5% to 10% when the specimens are water stored before the test.

2: The salt water uptake during freeze/thaw is higher than the salt water uptake during isothermal capillary suction.

3: The absorption is higher when freeze/thaw takes place in NaCl-solution than when it takes place in pure water.

4: The salt scaling is reduced when the air content is increased, or the water/cement ratio decreased. For the water/cement ratio 0.7, very high air contents are required.

5: The internal freeze stresses increase with increasing degree of saturation of salt water.

6: At constant degree of salt water saturation, the pessimum NaCl-concentration of pore water seems to be about 2.5%.

7 References

1. Petersson, O. (1991) The Chenical Effects on Cement Mortars of Solutions of Calcium Acetate and other Deicing Salts. Div. of Building Materials, Lund Institute of Technology, Report TVBM-3045.

2. Brown, F.P., Cady, P.D. (1975) Deicer Scaling Mechanisms in Concrete. American Concrete Institute, Special Publication SP-47, Detroit.

3. Fagerlund, G. (1975) Significance of Critical Degrees of Satruration at Freezing of Porous and Brittle Materials. American Concrete Institute, Special Publication SP-47, Detroit.

Part II

Parameters influencing the frost durability of concrete

5

La résistance à l'écaillage des BHP

R. GAGNÉ ET J. MARCHAND
Centre de recherche interuniversitaire sur le béton, Université Laval, Québec, Canada

Résumé
Les résultats d'une revue de la documentation scientifique et technique sur la résistance à l'écaillage dû aux cycles de gel et dégel en présence de sels fondants des bétons à haute performance (BHP) sont présentés. Les effets de différents facteurs, comme les paramètres de formulation et le type de mûrissement, sont passés en revue. Selon les données recueillies, la faible porosité des BHP contribue à assurer à la plupart des BHP une excellente résistance à l'écaillage. Selon certains travaux de recherche, il existerait même un rapport eau/liant critique en dessous duquel l'entraînement de l'air ne serait plus nécessaire pour protéger les BHP contre la détérioration par écaillage. Des recommandations pour assurer la résistance à l'écaillage des BHP sont formulées.
Mots-clefs: Écaillage, gel-dégel, durabilité, ciment, béton à haute performance.

1 Introduction

Dès la mise au point des premiers ciments Portland, la qualité du béton a constamment progressé, d'une part, grâce à la performance accrue des ciments Portland et, d'autre part, grâce à des techniques de formulation et de fabrication toujours en progrès. L'objectif principal était alors de faire progresser les propriétés mécaniques du béton en s'efforçant de réduire le plus possible le rapport eau/ciment (E/C) du mélange. Pour y parvenir, il fallait d'abord porter une attention particulière à la sélection du ciment et des granulats, pour ensuite optimiser le dosage des constituants de manière à diminuer au maximum la demande en eau du mélange. Cette technique a rapidement montré ses limites et, en l'absence des agents fluidifiants modernes, il était alors très difficile d'abaisser le rapport E/C en dessous de 0.40 et d'obtenir une résistance à la compression supérieure à 50 MPa.

Depuis l'apparition de plusieurs familles de fluidifiant, les propriétés mécaniques du béton ont rapidement effectué un bond considérable. À la fin des années soixante-dix, la résistance à la compression atteint déjà plus de 70 MPa et le qualificatif **béton à haute résistance** est alors de plus en plus utilisé. À cette époque, on cherchait surtout à fabriquer des bétons les plus résistants possible pour les utiliser dans des constructions de grande hauteur. On a cependant rapidement constaté que les bétons à haute résistance possédaient plusieurs particularités physico-chimiques susceptibles

Freeze-Thaw Durability of Concrete. Edited by J. Marchand, M. Pigeon and M. Setzer.
Published in 1997 by E & FN Spon, 2–6 Boundary Row, London SE1 8HN, UK.
ISBN 0 419 20000 2.

d'en faire des bétons beaucoup plus durables. De nombreux travaux de recherche ont permis de confirmer cette hypothèse et on a graduellement adopté le qualificatif plus général de **béton à haute performance** (BHP). Ce qualificatif souligne à la fois les meilleures caractéristiques mécaniques et rhéologiques de ces bétons, mais aussi leur plus grande durabilité.

En général, on reconnaît maintenant que les BHP sont des matériaux plus durables que les bétons usuels [1-3]. Cette amélioration de la performance concerne plusieurs aspects de la durabilité, notamment le comportement face aux cycles de gel-dégel et, plus particulièrement, la résistance à l'écaillage des surfaces exposées à l'action combinée des cycles de gel-dégel et des sels fondants. Même si, globalement, les BHP ont un meilleur comportement que les bétons usuels face aux sels fondants [4, 5], il apparaît que les gains en durabilité peuvent être très variables en fonction des paramètres de formulation du mélange (rapport eau/liant, ajouts minéraux, caractéristiques du réseau de bulles d'air). Par exemple, il existe des BHP qui sont tout à fait en mesure de résister à l'attaque des sels fondants sans la protection d'un bon réseau de bulles d'air [6-8]. Ce n'est cependant pas le cas pour tous les types de BHP. Ainsi, certains BHP ont absolument besoin d'être protégés par un bon réseau de bulles d'air pour résister à l'écaillage, malgré leur faible rapport E/C (0.30) [9, 20].

Pour l'ingénieur ou le spécialiste en technologie du béton, il est fondamental de connaître le plus précisément possible les caractéristiques minimales du réseau de bulles d'air entraîné qui permettront de garantir une bonne résistance du BHP face aux sels fondants. Le problème prend toute son importance puisqu'une augmentation de 1% du volume d'air entraîné peut diminuer la résistance à la compression de 4% à 5%. Connaissant les caractéristiques de formulation du BHP, l'ingénieur doit donc être en mesure de répondre à la question suivante: quelles sont les caractéristiques minimales du réseau de bulles d'air requises pour garantir un bon comportement du BHP face à l'attaque des sels fondants?

La réponse à cette question est rendue relativement complexe par le fait que le qualificatif BHP est assez large et peut s'appliquer à des bétons qui peuvent être très différents de par leur formulation, leurs propriétés physico-mécaniques ou leur mode de fabrication. Dans la documentation scientifique, il n'est pas surprenant de constater que le comportement des BHP face aux sels fondants se révèle relativement variable, ce qui vient compliquer d'autant le travail de l'ingénieur. L'objet de cette publication est de présenter une synthèse des études portant sur l'évaluation de la résistance à l'écaillage des BHP pour en arriver à proposer aux concepteurs d'ouvrages des recommandations et des limites pratiques concernant la formulation des BHP soumis à l'action des sels fondants.

2   Précision du cadre de la synthèse bibliographique

Avant d'entreprendre une synthèse des études portant sur la résistance à l'écaillage des bétons à haute performance, il convient d'abord d'en préciser certains termes. Cette synthèse a été menée en considérant, comme BHP, tous les bétons à base de ciment Portland dont la résistance moyenne à la compression à 28 jours est comprise entre 60 et 100 MPa. Conformément à la pratique usuelle, les bétons dont la résistance moyenne à la compression à 28 jours est supérieure à 100 MPa ont été classés dans la catégorie BTHP (bétons à très haute performance)[10]. Cette distinction est basée sur le fait qu'on ne peut atteindre ce très haut niveau de résistance qu'avec des matériaux de très haute qualité et en utilisant un rapport eau/liant (E/L) encore plus faible (généralement de l'ordre de 0.25 et moins). Le niveau de performance de cette catégorie de béton est significativement plus élevé que celui des BHP, ce qui justifie la création d'une famille qui lui soit propre. Dans le cas particulier de cette publication, nous avons choisi d'élargir un peu plus la notion de BHP en y incluant aussi tous les bétons dont le rapport E/L est inférieur à 0.30, même si leur résistance à 28 jours est inférieure à 60 MPa. Ce choix est basé sur le fait que le rapport E/L est le principal paramètre qui gouverne la performance du béton. Il permet d'inclure des bétons contenant une forte proportion d'adjuvants minéraux (fumée de silice, cendres volantes, laitiers) qui, malgré un rapport E/L faible, n'ont pas toujours une très haute résistance à 28 jours.

Nous avons également élargi cette revue de la documentation à tous les produits de béton qui sont fabriqués en ayant recours à la technologie des bétons secs. Bien qu'ils ne soient généralement pas considérés comme des BHP, les bétons secs partagent avec ces derniers plusieurs similitudes. Tout comme les BHP classiques, les bétons secs sont fabriqués avec des rapports E/L faibles. Selon le type de produit et l'application envisagée, le rapport E/L des bétons secs varie généralement de 0.22 à 0.35. Tout comme les BHP classiques, les bétons secs peuvent également atteindre des résistances mécaniques très respectables. Bien que ce ne soit jamais le cas pour les bétons compactés au rouleau utilisés dans la construction de barrage (qui sont généralement fabriqués avec de très faibles quantités de liant), la résistance à la compression des bétons secs dépasse souvent les 50 MPa. Il n'est pas rare de voir des ouvrages de bétons secs atteindre des résistances à la compression supérieures à 70 MPa.

L'écaillage est une détérioration qui survient le plus souvent lorsque le béton est exposé à l'action combinée des cycles de gel-dégel et des sels fondants (généralement NaCl ou $CaCl_2$) [11, 4]. Il s'agit du type de destruction par le gel le plus fréquemment observé sur les structures de béton en contact avec des sels fondants. Essentiellement, c'est un phénomène de surface dont le mécanisme de destruction est différent de celui qui peut causer la fissuration interne des bétons exposés à des cycles de gel-dégel dans des conditions voisines de la saturation [11]. Les caractéristiques minimales du réseau de bulles d'air requises pour protéger un béton contre chacun de ces deux types de détérioration peuvent être très différentes, car les mécanismes de destruction ne sont pas les mêmes [11].

Cette synthèse des études sur la résistance à l'écaillage des BHP est exclusivement basée sur des résultats de laboratoire. On peut simuler et évaluer en laboratoire la résistance à l'écaillage du béton en utilisant des essais du genre Ponding Method. Ce type d'essai consiste à soumettre à des cycles de gel-dégel journaliers, une surface horizontale de béton constamment en contact avec une solution contenant des sels fondants. Pour accélérer la vitesse de destruction, on utilise les concentrations les plus agressives qui, en fonction du type de fondant, sont généralement comprises entre 3% à 5% en masse. Les essais les plus fréquemment utilisés sont l'essai ASTM C 672 [12] et l'essai suédois SS 13 72 44 [13] (méthode Borås). Le principe de ces deux essais est le même, mais l'essai suédois peut être considéré comme une version améliorée et mieux contrôlée de l'essai ASTM C 672. Dans les deux cas, seule la surface horizontale supérieure d'une éprouvette de béton est exposée à la solution saline. Les éprouvettes sont ainsi exposées à une cinquantaine de cycles de gel-dégel d'une durée de 24 heures. Les principales différences entre ces essais sont les températures minimale et maximale, le nombre minimal de cycles, le préconditionnement des éprouvettes avant l'essai et l'isolation thermique des parois des éprouvettes (Tabl. 1).

Même s'il existe plusieurs différences entre ces procédures, on peut considérer que les résultats de l'essai ASTM et de l'essai suédois sont comparables dans la mesure où ils sont utilisés pour estimer la performance relative de différents types de BHP soumis à l'attaque des sels fondants. Pour l'essai ASTM, il n'existe pas de critère d'acceptabilité dans le texte de la norme. Cependant, conformément à la pratique, on peut estimer la résistance à l'écaillage à partir de la masse de débris par m^2 de surface exposée après 50 cycles (Tabl. 2). Dans le cas de l'essai suédois, les critères s'appliquent à la masse des débris après 56 cycles (Tabl. 2). À l'origine, ces critères ont été proposés pour des surfaces sciées, mais on les utilise couramment pour l'évaluation des surfaces moulées.

Tableau 1. Principales caractéristiques des essais ASTM C 672 et SS 13 72 44

Caractéristiques	ASTM C 672	SS 13 72 44
Nombre de cycles	50	56
Surface minimale exposée	2 éprouvettes de 465 cm^2 chacune	2 éprouvettes de 108 cm^2 chacune
Préconditionnement	14 d à 50% H.R.	7 d à 50% H.R. + 3 d de saturation à l'eau
Type de sel fondant	Solution à 4% de CaCl$_2$	Solution à 3% de NaCl
Isolation de l'éprouvette	Aucune	Tous les côtés et la surface inférieure
Protection contre l'évaporation	Aucune	Membrane étanche sur le dessus
Température min. du cycle	-18° C ($\pm$ 2° C)	-18° C ($\pm$ 2° C)
Température max. du cycle	18° C ($\pm$ 2° C)	20° C ($\pm$ 5° C)
Nombre d'heures de gel	16 h à 18 h	environ 16 h *
Nombre d'heures de dégel	6 h à 8 h	environ 8 h *
Fréquence des mesures	À tous les 5 cycles	7, 14, 28, 42 et 56 cycles

* La norme spécifie une plage de variation des températures pour un cycle complet

Tableau 2. Critères pour l'évaluation de la résistance à l'écaillage

	Masse des débris (kg/m^2)	
Résistance à l'écaillage	ASTM C 672 (50 cycles)	SS 13 72 44 (56 cycles)
Très bon	< 0.10	< 0.10
Bon	< 0.75	< 0.50*
Acceptable	< 1.0	< 1.0*

* Pour ces catégories, on doit en outre satisfaire à une exigence supplémentaire qui spécifie que le rapport entre la masse des débris à 56 cycles et la masse des débris à 28 cycles devrait être inférieur à 2.

3. Revue des résultats publiés

Le comportement à l'écaillage des BHP est gouverné par de nombreux facteurs qui peuvent être regroupés en trois familles. Une première famille comprend l'ensemble des paramètres reliés à la formulation du mélange. Parmi ceux-ci, on retrouve notamment le rapport E/L, les propriétés du ciment, le type d'ajout minéral et les caractéristiques du réseau de bulles d'air. La seconde famille regroupe les facteurs décrivant les conditions de mûrissement précédant la première exposition aux sels fondants ou aux cycles de gel. Elle regroupe des paramètres tels que la durée, la température et les conditions d'humidité relative de la cure. La troisième famille regroupe les facteurs associés aux conditions d'exposition ou au vieillissement du matériau (âge, état de fissuration, influence du degré de séchage). Pour faciliter le travail de synthèse, la revue bibliographique sur le comportement à l'écaillage des BHP a été organisée en fonction de ces trois grandes familles de paramètres.

3.1 Influence des paramètres de formulation

On a déjà mentionné que les paramètres de formulation des BHP sont nombreux et qu'ils peuvent, individuellement, adopter des valeurs comprises dans une plage de variation relativement large. Pour simplifier la présentation et l'analyse des résultats disponibles dans la documentation, nous avons regroupé les BHP selon 5 familles définies en fonction du type de liant utilisé ou du procédé de fabrication:

*   béton sans ajout (100% ciment Portland);
*   béton avec fumée de silice (5 à 10% de remplacement du ciment);
*   béton avec cendres volantes (jusqu'à 60% de remplacement du ciment);
*   béton avec laitiers de haut fourneau;
*   béton sec.

Pour chacune de ces 5 grandes familles de BHP, la performance relative face aux sels fondants a principalement été comparée en fonction du rapport E/L du mélange et non pas en fonction de la résistance à la compression. Ce choix découle du fait que les mécanismes de destruction par écaillage sont surtout contrôlés par les propriétés physico-chimiques de la pâte de ciment durci. Le rapport E/L apparaît le paramètre de formulation le plus étroitement relié aux caractéristiques et à la qualité générale de la pâte. La résistance à la compression est influencée par un plus grand nombre de paramètres qui ne sont pas toujours fonction des propriétés de la pâte. Par exemple, en optimisant le mélange granulaire, il est possible de produire des BHP avec une résistance à la compression de près de 100 MPa avec un rapport E/C relativement élevé de 0.36 [34]. La durabilité des BHP fabriqués selon cette approche (E/L relativement élevé) est souvent assez décevante.

3.1.1 Béton sans ajout

Les données disponibles dans la documentation montrent clairement que le rapport E/C est un paramètre fondamental contrôlant la résistance à l'écaillage. La Fig. 1, qui regroupe des résultats de plusieurs chercheurs, présente la masse des débris après 50 ou 56 cycles de gel pour plusieurs bétons en fonction du rapport E/C et de la présence ou non d'un réseau de bulles d'air entraîné [8, 9, 14-20]. Dans tous les cas, il s'agit de bétons sans ajout fabriqués avec un ciment Portland ordinaire. La Fig. 2, similaire à la Fig. 1, présente des résultats obtenus après 100 ou 150 cycles de gel [8, 9, 14, 20].

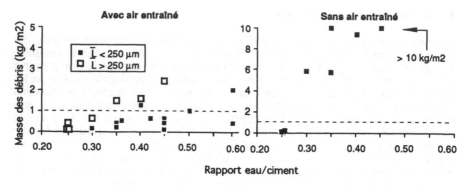

Fig. 1 Masse des débris, après 50-56 cycles, en fonction du rapport E/C et des caractéristiques des bulles d'air. Bétons avec ciment Portland ordinaire sans ajout [adapté des références 8, 9, 14-20]

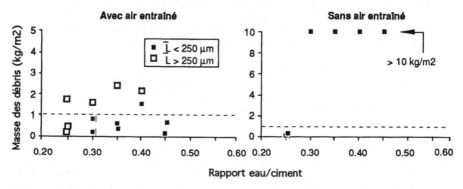

Fig. 2 Masse des débris, après 100 ou 150 cycles, en fonction du rapport E/C et des caractéristiques des bulles d'air. Bétons avec ciment Portland ordinaire sans ajout [adapté des références 8, 9, 14, 20]

La résistance à l'écaillage des bétons à air entraîné ayant un facteur d'espacement inférieur à 250 µm ne pose évidemment pas de problème en autant que le rapport E/C demeure inférieur à 0.50 (Fig. 1). Cependant, si le réseau de bulles d'air entraîné est de mauvaise qualité (facteur d'espacement > 250 µm), il est possible d'obtenir une résistance à l'écaillage inacceptable en dépit d'un rapport E/C aussi bas que 0.35. Signalons que, dans certains cas, l'air entraîné peut diminuer légèrement la résistance à l'écaillage des BHP fabriqués avec un très faible rapport E/C (0.25) [20, 21]. Cette question sera traitée plus en détail un peu plus loin dans cette section.

Les données concernant la résistance à l'écaillage des bétons sans air entraîné sont beaucoup moins nombreuses, mais il est néanmoins possible d'identifier une tendance générale. La résistance à l'écaillage des bétons sans air entraîné est généralement très mauvaise si le rapport E/C est de 0.30 et plus (Fig. 1). On note une certaine amélioration de la résistance à l'écaillage pour des rapports E/C de 0.30 ou 0.35, mais, après 50 cycles, la masse des débris est encore bien au-delà du critère d'acceptabilité de 1 kg/m². Ce n'est qu'en abaissant le rapport E/C à des valeurs très faibles, soit aux environs de 0.25, que l'on parvient à produire des BHP en mesure de résister aux sels fondants sans la protection d'un réseau de bulles d'air entraîné. Les tendances observées après 100 ou 150 cycles sont similaires à celles observées après 50-56 cycles sauf que, bien entendu, les destructions finales sont plus importantes.

Les propriétés physiques et chimiques des ciments (finesse, teneur en C_3A ou C_3S) peuvent influencer la résistance à l'écaillage des BHP. Les résultats de Gagné et coll. [6] ont démontré que des BHP fabriqués avec un rapport E/C de 0.30 et un ciment de type 30[1] pouvaient avoir une très bonne résistance à l'écaillage sans qu'il soit nécessaire de les protéger par un réseau de bulles d'air entraîné (masse des débris inférieure à 0.2 kg/m² après 150 cycles pour un facteur d'espacement de 950 μm). Ces performances sont d'autant plus satisfaisantes que ces résultats ont été obtenus après seulement 1 et 3 jours de mûrissement scellé. Des études sont en cours afin d'identifier et de mieux comprendre comment certaines propriétés du ciment peuvent influencer la résistance à l'écaillage des BHP.

Quelques études ont déjà permis de constater que l'air entraîné pouvait parfois diminuer légèrement la résistance à l'écaillage des BHP [8, 20, 21]. Ce phénomène s'observe généralement lorsque l'on entraîne de l'air dans des bétons qui n'ont pas besoin de la protection d'un bon facteur d'espacement pour être très résistants à l'écaillage. Deux cas typiques sont présentés à la Fig. 3 [6, 8]. Il semble que cette destruction supplémentaire serait provoquée par la saturation progressive de très nombreuses petites bulles d'air près de la surface exposée [20, 22]. Lorsque le degré de saturation des bulles devient très élevé, la formation de glace dans les bulles peut provoquer des pressions internes qui font éclater la pâte au voisinage de la surface. Il est important de signaler que, dans presque tous les cas, l'augmentation de la sensibilité à l'écaillage demeure faible et la masse des débris est généralement très inférieure à 1 kg/m² après 50 ou 56 cycles.

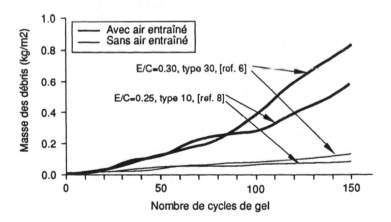

Fig. 3 Influence typique de l'air entraîné sur la résistance à l'écaillage de bétons qui n'ont pas besoin de la protection d'un bon facteur d'espacement pour résister à l'écaillage [adapté des références 6 et 8]

---

[1] Ciment à haute résistance initiale dont les caractéristiques physico-chimiques sont similaires à celles d'un ciment ASTM Type III ou un CPA HPR.

D'un point de vue pratique, le comportement à l'écaillage des BHP fabriqués avec un ciment Portland sans ajout peut se résumer de la manière suivante : ces BHP ont une bonne résistance à l'écaillage sans air entraîné en autant que le rapport E/C soit d'environ 0.25. Il s'agit bien alors de **bétons à très haute performance** (BTHP). Pour des rapports E/C supérieurs à 0.25, un réseau de bulles d'air entraîné demeure nécessaire et ce n'est plus le rapport E/C, mais plutôt les caractéristiques du réseau de bulles d'air qui exercent la plus grande influence sur le comportement du béton face aux sels fondants. L'emploi d'un ciment plus performant (haute résistance initiale) peut permettre de déplacer ce rapport E/C limite aux environs de 0.30. Des essais spécifiques en laboratoire devraient être alors utilisés pour valider la résistance à l'écaillage. La présence d'un fort volume d'air entraîné dans certains BHP ou BTHP peut parfois augmenter légèrement leur sensibilité à l'écaillage sans pour autant compromettre leur performance générale.

3.1.2 Béton avec fumée de silice

Un relevé de plusieurs résultats publiés dans la documentation permet de mettre en évidence la relation entre le rapport E/L et la résistance à l'écaillage des bétons avec fumée de silice. La figure 4 présente un ensemble de résultats d'écaillage obtenus à partir de bétons avec fumée de silice avec et sans air entraîné [6, 7, 14, 20]. Tous ces résultats ont été obtenus après 100 cycles de gel. Deux principaux types de ciment Portland ont été utilisés : un ciment Porltand ordinaire (CSA type 10 ou ASTM type I) et un ciment Portland à performance améliorée (CSA type 30 ou ASTM Type III et un P30 4A norvégien). Les taux de remplacement du ciment varient de 5 à 10%. Il n'existe que très peu de résultats concernant la résistance à l'écaillage de bétons fabriqués avec des taux de remplacement plus élevés puisque l'on reconnaît maintenant qu'un taux de remplacement de 10% constitue une limite supérieure [4].

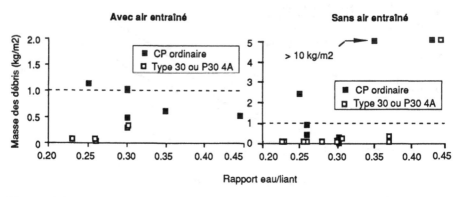

Fig. 4   Masse des débris, après 100 cycles, en fonction du rapport E/L de bétons contenant de 5 à 10% de fumée de silice avec ou sans air entraîné [adapté des références 6, 7, 14, 20]

On constate d'abord que la résistance à l'écaillage des BHP à air entraîné ne pose évidemment pas de problème quel que soit le rapport E/L du mélange. Deux résultats sont néanmoins légèrement supérieurs à la limite de 1 kg/m^2, mais il faut rappeler qu'il s'agit, dans tous les cas, de destructions mesurées après 100 et non pas 50 cycles de gel. Le choix d'un ciment plus performant (du type CSA type 30 ou norvégien P30 4A) produit généralement une amélioration sensible de la résistance à l'écaillage. Comme dans le cas des BHP sans ajout, de nombreux résultats démontrent qu'il est possible de produire des BHP avec fumée silice qui aient une très bonne résistance à l'écaillage sans qu'il ne soit nécessaire de le protéger par un réseau de bulles d'air entraîné. Il faut cependant abaisser le rapport E/L jusqu'à des valeurs relativement faibles, de l'ordre de 0.30 et moins. Un seul résultat obtenu, par Marchand et coll. [4], fait état d'une masse de débris supérieure à 1 kg/m^2 pour un rapport E/L de 0.25. On peut cependant considérer qu'il s'agit d'une performance acceptable puisque, après 50 cycles, la masse des débris était tout juste au-dessus de la limite supérieure (1.04 kg/m^2). On remarque encore une fois que l'utilisation d'un ciment Portland plus performant permet de produire des BHP significativement plus résistants à l'écaillage.

L'effet légèrement néfaste des forts volumes d'air entraîné sur la résistance à l'écaillage des BHP avec fumée de silice a aussi été signalé dans la documentation [6, 20]. L'influence sur la performance générale des BHP soumis à l'attaque des sels fondants demeure cependant mineure.

Il arrive que la destruction par écaillage des bétons avec fumée de silice ne progresse pas de façon linéaire en fonction du nombre de cycles [14, 19]. Dans certains cas, la masse des débris est très faible après les 50 premiers cycles, mais on peut observer une progression beaucoup plus rapide de la destruction pendant les 50 ou 100 cycles suivants. L'augmentation graduelle du degré de saturation au-delà d'un seuil critique pourrait expliquer ce phénomène [14]. Selon cette hypothèse, ce type de comportement risque peu de se produire en pratique puisqu'il est rare qu'une surface de béton soumise à l'action des sels fondants demeure continuellement immergée sous une pellicule d'eau pendant plusieurs mois consécutifs.

Même si les études portant sur la résistance à l'écaillage des BHP avec fumée de silice sont encore peu nombreuses, on peut néanmoins en tirer quelques recommandations. Il est tout à fait possible de fabriquer des BHP avec fumée de silice qui soient parfaitement durables à l'écaillage sans qu'il ne soit nécessaire d'y entraîner un réseau de bulles d'air. Nous suggérons pour cela de limiter le taux de remplacement à 10% et d'utiliser un rapport E/L de l'ordre de 0,30. De nombreux BHP avec fumée de silice sont, par conséquent, en mesure de résister aux sels fondants sans air entraîné. Il est trop tôt pour affirmer que la fumée de silice améliore systématiquement la résistance à l'écaillage des BHP, même si plusieurs très bonnes performances ont été obtenues avec un rapport E/L de 0,30. Les résultats concernant l'influence de la fumée de silice sur la résistance à l'écaillage du béton sont encore trop contradictoires [18, 23-25]. Seul l'emploi d'un ciment Portland plus performant (à haute performance ou à haute résistance initiale) peut vraisemblablement permettre d'utiliser un rapport E/L de 0.35 et moins sans qu'il ne soit nécessaire d'ajouter un agent entraîneur d'air pour assurer la protection contre les sels fondants.

3.1.3 Béton avec cendres volantes

Le comportement face à l'action combinée des cycles de gel-dégel et des sels fondants est un des points faibles de la durabilité des bétons de résistance normale contenant des cendres volantes (CV) en remplacement du ciment. On retrouve de nombreuses mentions dans la documentation qui signalent que les bétons avec CV (type F ou C) sont plus sensibles à l'écaillage que des bétons de référence ne contenant pas de cendre [26, 30-33]. Les performances sont, en général, assez variables d'une étude à l'autre, probablement en raison de la nature très variable des caractéristiques physico-chimiques des CV. Même si les bétons avec CV se révèlent souvent plus sensibles à l'écaillage, il est néanmoins possible d'obtenir des performances tout à fait satisfaisantes en s'assurant de produire un facteur d'espacement des bulles d'air adéquat (facteur d'espacement < 250 µm), en limitant les taux de remplacement à des valeurs de 30% et moins et en fournissant des conditions de mûrissement favorables [4, 30-33].

À notre connaissance, parmi les quelques publications portant sur la résistance à l'écaillage des BHP avec CV, il n'en existe qu'une seule qui fasse état d'une bonne résistance à l'écaillage de BHP avec CV sans air entraîné [28]. Ces résultats ont été obtenus avec des rapports E/L très faibles (0.22 et 0.25) où 25% du liant (un ciment de type 10 avec fumée de silice et un ciment de type 30) avait été remplacé par une CV de type F. La plupart des autres publications tendent plutôt à démontrer que le comportement à l'écaillage des BHP contenant des CV n'est pas très différent de celui des bétons avec CV de résistance normale. En effet, quel que soit le rapport E/L utilisé, la présence d'un bon réseau de bulles d'air entraîné apparaît une condition essentielle pour obtenir une résistance à l'écaillage acceptable [17, 26-28].

Les résultats de la Fig. 5 permettent d'illustrer la relation entre la résistance à l'écaillage de bétons avec CV en fonction du rapport E/L et du taux de remplacement du ciment (CV type F). Tous les bétons sont à air entraîné et les masses des débris ont été obtenues après 50 ou 56 cycles de gel-dégel. La période de mûrissement humide varie entre 14 et 28 jours.

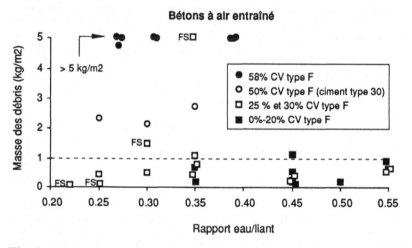

Fig. 5   Masse des débris, après 50-56 cycles, en fonction du rapport E/L de bétons à air entraîné contenant des CV de type F. (FS: ciment avec fumée de silice) [adapté des références 17, 26-28]

Pour un taux de remplacement de 30% ou moins, il est généralement possible d'obtenir une résistance à l'écaillage acceptable, quel que soit le rapport E/L, en autant que le béton possède un bon réseau de bulles d'air. Il faut cependant souligner que Reid et coll. [28] ont obtenu deux mauvaises performances avec un taux de remplacement de 25% (ciment avec fumée de silice) et pour des rapports E/L de 0.30 et 0.35. Ces écarts de performance sont assez fréquents dans le cas des bétons avec CV et ils sont, pour une large part, expliqués par la nature très variable des cendres.

Les taux de remplacement de 50% ou de 60% sont à éviter lorsque le béton risque d'être en contact avec des sels fondants. Il est important de souligner que, pour ces forts contenus en CV, même un bon facteur d'espacement (facteur d'espacement < 250 µm) et un rapport E/L de 0.30 et moins ne permettent généralement pas d'obtenir une bonne performance face aux sels fondants.

La résistance à l'écaillage relativement décevante des BHP avec CV est quelquefois expliquée par la moins grande réactivité des CV qui retarde le niveau de maturité du béton. Cette hypothèse tend graduellement à être écartée, car des essais récents semblent démontrer que la prolongation du mûrissement (jusqu'à 28 jours) avant la première exposition aux sels n'a pas d'effet significatif sur la performance à l'écaillage des bétons avec CV [26].

Plusieurs aspects du comportement à l'écaillage des bétons avec CV sont encore mal compris, dont l'influence des propriétés physico-chimiques des cendres. On doit donc être prudent lorsque l'on envisage d'utiliser un béton contenant des CV dans des structures fortement exposées à l'action des sels fondants. Pour le moment, on ne peut donc que recommander de toujours prévoir un bon facteur d'espacement, quel que soit le rapport E/L ou le niveau de résistance du béton. Il semble aussi prudent de limiter les taux de remplacement à un maximum de 30%.

3.1.4 Béton avec laitiers de haut fourneau

Les résultats concernant la résistance à l'écaillage des bétons avec laitiers de haut fourneau sont relativement rares et souvent contradictoires. Fagerlund mentionne que la résistance à l'écaillage des bétons augmente avec le contenu en laitier [17, 36] alors que Virtanen signale plutôt que la sensibilité à l'écaillage augmente avec la proportion de ciment remplacé par des laitiers [24].

Les études portant spécifiquement sur le comportement à l'écaillage des BHP avec laitiers de haut fourneau sont encore plus rares. Baalbaki [38] a fabriqué un BHP sans air entraîné de 120 MPa (91 d) en utilisant un rapport E/L de 0.31 et un liant à base de ciment Portland contenant 40% de laitier et 10% de fumée de silice. Des essais d'écaillage non publiés ont montré que, après 3 jours de mûrissement scellé et 28 jours de séchage, la masse des débris a dépassé 2.5 kg/m^2 après 50 cycles de gel. Malgré des performances mécaniques remarquables, ces bétons sans air entraîné peuvent donc se révéler relativement sensibles à l'attaque des sels fondants.

Quelques résultats concernent la résistance à l'écaillage de bétons dont le liant est constitué à 100% de laitiers activés par des alcalis [35]. Il s'agit de plusieurs bétons, avec et sans air entraîné, fabriqués avec des rapports E/L compris entre 0.50 et 0.32 dont la résistance à la compression à 100 jours est comprise entre 50 et 80 MPa. Les essais d'écaillage ont débuté après 7 jours de mûrissement humide et environ 100 jours de séchage. Pour les rapports E/L de 0.42 et 0.50, on a mesuré une mauvaise résistance à l'écaillage quelle que soit la teneur en air (masse des débris > 4 kg/m^2 après 56 cycles). La résistance à l'écaillage des bétons avec un rapport E/L de 0.32 a tout juste été acceptable ($\approx$ 1 kg/m^2 après 56 cycles) et on n'a pas noté d'effet bénéfique de la teneur en air sur la résistance à l'écaillage.

La revue des résultats publiés jusqu'à présent démontre que le comportement face aux sels fondants des bétons ordinaires ou des BHP contenant des laitiers de haut fourneau a été peu étudié et est encore très mal connu. Par conséquent, l'utilisation de BHP contenant des laitiers de haut fourneau dans des structures soumises à l'action des sels fondants doit se faire avec grande prudence. Il est encore trop tôt pour formuler des recommandations précises, mais il apparaît prudent d'éviter les taux de remplacement élevés et il convient de s'assurer de la présence d'un bon réseau de bulles d'air entraîné.

3.1.5 Béton sec

On distingue généralement trois grandes catégories de bétons secs : les bétons sans affaissement (ou zero-slump concretes), les pavés de béton et les bétons compactés au rouleau. Bien qu'ils soient quelquefois utilisés pour la construction d'ouvrages en chantier, les bétons sans affaissement servent principalement à la production de pièces préfabriquées en usine. Par exemple, on retrouve dans cette catégorie les bordures de rue et les barrières de sécurité de type New-Jersey. Tout comme les pavés de béton, les bétons sans affaissement sont moulés, la consolidation du mélange se faisant par vibration et compactage. Les bétons sans affaissement se distinguent cependant des pavés de béton à deux points de vue : ils sont généralement fabriqués avec des rapports E/L plus élevés et avec des granulats dont la dimension maximale est plus importante. Comme leur nom l'indique, les bétons compactés au rouleau sont mis en place à l'aide d'un rouleau vibrateur. Les bétons compactés au rouleau sont principalement utilisés dans la construction de barrages et de revêtements routiers. Les caractéristiques des mélanges employés diffèrent considérablement selon le type envisagé d'application. Ainsi, les bétons de barrage sont fabriqués avec des teneurs en liant nettement plus faibles et avec de très gros granulats. Les bétons compactés au rouleau servant à la construction de barrages ne sont pas inclus dans cette revue.

Étant donné leur maniabilité réduite et le peu d'eau ajoutée au mélange, il est généralement très difficile d'entraîner des bulles d'air dans les bétons secs. Du point de vue de la résistance à l'écaillage, ceci constitue un problème majeur. Comme il a été déjà mentionné, même pour les bétons de faible rapport E/L, l'air entraîné constitue souvent une condition nécessaire pour assurer une bonne résistance à l'écaillage.

Pour cette raison, plusieurs chercheurs se sont intéressés à l'entraînement de l'air dans les bétons secs au cours des dernières années. Certaines études ont ainsi démontré qu'il était possible d'entraîner des bulles d'air dans les mélanges de béton sans affaissement lorsque des dosages très importants en agent entraîneur d'air étaient utilisés [39-41]. À la suite de ces études, il apparaît que les rapports E/L relativement plus élevés des bétons sans affaissement facilitent l'action de l'agent entraîneur d'air.

À notre connaissance, il existe très peu d'études sur l'entraînement de l'air dans les pavés de béton. Les seules données publiées sur le sujet démontrent cependant qu'il est possible d'entraîner de l'air dans ces mélanges [42, 43]. Dans le cas des pavés de béton, il semble que l'entraînement de l'air soit directement fonction du type d'agent entraîneur d'air utilisé et de son dosage. Curieusement, les résultats de ces études indiquent que l'emploi d'un agent entraîneur d'air contribue à produire principalement des bulles d'air de très petites dimensions. Ces microbulles, invisibles au microscope optique, sont uniquement détectables au microscope électronique à balayage.

L'entraînement de l'air dans les bétons compactés au rouleau fait toujours aujourd'hui l'objet d'une controverse. Alors que certaines études semblent indiquer qu'il est possible de produire des bulles d'air sphériques dans ces mélanges [44-46], d'autres tendent plutôt à démontrer que l'entraînement de l'air dans les bétons compactés au rouleau est une opération extrêmement difficile [47-49]. Il est important de souligner que toutes les études qui indiquent que l'ajout d'un agent entraîneur d'air peut avoir un effet bénéfique ont été réalisées en laboratoire alors que les autres ont été réalisées en chantier. Cette distinction est assez importante. Une étude récente sur l'entraînement de l'air dans les bétons secs a, en effet, clairement démontré que l'efficacité de l'agent entraîneur d'air ajouté à ce type de mélange est directement fonction de la séquence de malaxage et du type de malaxeur utilisé [43]. Toutes les études de laboratoire ont été réalisées avec des malaxeurs nettement plus efficaces que ceux que l'on retrouve en pratique. De façon générale, il apparaît que l'entraînement de l'air dans les bétons secs est une opération très complexe qui nécessite une bonne connaissance des matériaux utilisés et l'emploi d'un malaxeur très puissant.

Étant donné que les bétons secs sont principalement utilisés dans la fabrication d'ouvrages exposés à l'action combinée des cycles de gel-dégel et des sels fondants, leur résistance à l'écaillage a fait l'objet d'un certain nombre d'études au cours des dernières années. Globalement, il semble que les bétons secs soient plus susceptibles de se détériorer par écaillage que des bétons conventionnels de même rapport E/L. Bien que cette plus grande susceptibilité soit, en grande partie, liée aux problèmes d'entraînement de l'air, il apparaît que les plus faibles résistances à l'écaillage des bétons secs soient également reliées à la microstructure de ces bétons [50]. En effet, une étude récente a démontré que les bétons secs ont tendance à avoir une pâte de ciment nettement plus hétérogène que les bétons conventionnels [45]. Cette hétérogénéité serait principalement attribuable à une mauvaise dispersion de l'eau de gâchage lors du malaxage.

Bien que les bétons secs apparaissent, dans l'ensemble, moins résistants à l'écaillage, plusieurs études ont cependant indiqué qu'il est tout de même possible de produire des ouvrages durables. En ce qui concerne les bétons sans affaissement et les pavés de béton, certaines études tendent à démontrer que de très faibles teneurs en air (de l'ordre de 1.5 à 2%) permettent généralement de protéger adéquatement ces bétons contre l'écaillage [41, 50]. Dans l'éventualité où l'ajout d'un agent entraîneur d'air se révèle tout à fait inefficace, il semble que la réduction du rapport E/L en dessous d'une certaine limite (≈ 0.30) permette de produire des bétons durables [41, 50, 51]. Dans la pratique, cette réduction du rapport E/L s'accompagne généralement d'une augmentation de la teneur en liant du mélange de manière à maintenir la consistance de celui-ci [52, 53].

Dans l'ensemble, il semble que, pour les bétons sans affaissement et les pavés de béton, l'emploi d'ajouts minéraux, comme les fumées de silice et les CV, n'ait pas d'effet significatif sur le comportement à l'écaillage des ouvrages [41, 50]. De plus, dans le cas des pavés de béton, la plupart des producteurs hésitent à utiliser de grandes quantités d'ajouts minéraux qui pourraient éventuellement améliorer la durabilité de ces bétons. Les ajouts minéraux tendent en effet à affecter la couleur des pavés en atténuant l'action des colorants ajoutés au mélange.

Dans le cas des revêtements routiers en béton compacté au rouleau, il semble qu'un des moyens les plus sûrs de produire des mélanges durables, sans ajout d'air entraîné, soit d'employer de la fumée de silice [45, 48, 49, 54, 55]. En plus d'augmenter la compacité de la pâte de ciment, l'ajout de fumée de silice permet également d'améliorer l'homogénéité de celle-ci en favorisant une bonne dispersion de l'eau de gâchage lors du malaxage [51]. La fumée de silice contribue ainsi à améliorer considérablement la résistance mécanique des bétons compactés au rouleau et à réduire de manière significative leur perméabilité [50, 56]. Il en résulte des revêtements routiers plus aptes à résister à l'action combinée des cycles de gel-dégel et des sels fondants.

Jusqu'à ce jour, les résultats concernant l'effet des CV sur la résistance à l'écaillage des revêtements de béton compacté au rouleau semblent plus ou moins contradictoires. Alors que certaines études indiquent qu'il est possible de produire des ouvrages résistant à l'écaillage en employant des CV [48, 50], d'autres études démontrent plutôt que les CV ont un effet négatif sur la durabilité des revêtements de béton compacté au rouleau [49, 57]. L'utilisation des CV dans ces bétons devrait donc être considérée avec une certaine prudence.

3.2 Influence du mûrissement

Comme il a été souligné à la section précédente, la réduction de la porosité capillaire contribue généralement à améliorer, de manière significative, la résistance à l'écaillage du béton. La réduction de l'espace capillaire permet en effet d'augmenter les propriétés mécaniques de la pâte de ciment et de limiter la quantité d'eau gelable qu'elle contient. Tout comme l'abaissement du rapport E/L et l'utilisation de certains ajouts minéraux, un des moyens de réduire la porosité capillaire du béton est de lui fournir un mûrissement adéquat.

Il est maintenant bien connu que les conditions de mûrissement affectent principalement les couches superficielles d'un ouvrage. C'est d'ailleurs pourquoi, de toutes les propriétés du béton, la résistance à l'écaillage est probablement celle qui est le plus influencée par les conditions de cure. L'influence du mûrissement sur la résistance à l'écaillage des BHP fait l'objet de cette section.

3.2.1 Influence du type de mûrissement

Il a été démontré que l'hydratation d'une pâte de ciment s'arrête si celle-ci est conservée dans une atmosphère dont l'humidité relative est inférieure à 80% [58, 59]. En laboratoire, pour maintenir l'humidité relative du béton au-dessus de cette limite, la pratique courante consiste généralement à immerger les éprouvettes de béton dans un bain d'eau saturée de chaux ou à les entreposer dans une chambre gardée à 100% d'humidité relative. En pratique, on a plutôt tendance à recouvrir les ouvrages de jute humide ou à appliquer, en surface du béton, une membrane de mûrissement. Celle-ci se distingue des scellants, que l'on utilise pour protéger certains ouvrages, à plusieurs points de vue. Alors que la membrane de mûrissement limite l'évaporation de l'eau en surface, les scellants sont plutôt appliqués pour empêcher la pénétration de l'eau et des ions chlore dans l'ouvrage. De plus, n'étant pas coupe-vapeur, les scellants n'empêchent nullement l'eau de s'évaporer du béton.

Curieusement, on retrouve très peu d'information dans la documentation sur l'influence du type de mûrissement sur la résistance à l'écaillage du béton en général et des BHP en particulier. En ce qui concerne les bétons ordinaires, le type de mûrissement ne semble pas avoir une influence très nette sur le comportement à l'écaillage. Alors que, dans certains cas, le type de mûrissement n'apparaît pas avoir d'effet significatif [32, 60], dans d'autres, le mûrissement à la membrane semble donner de meilleurs résultats [26]. À notre connaissance, seuls Bilodeau et coll. [26] se sont intéressés à l'influence du type de mûrissement sur la résistance à l'écaillage des BHP. Leurs résultats démontrent que l'application d'une membrane tend à augmenter légèrement la résistance des BHP, et tout particulièrement celle des mélanges contenant des CV. Plusieurs études indiquent également que le mûrissement à la membrane permet d'améliorer, de façon considérable, la résistance à l'écaillage des revêtements de béton compacté au rouleau [48-50].

L'ingénieur ou le spécialiste en technologie du béton qui doit choisir un type de mûrissement pour un ouvrage exposé à l'action combinée des cycles de gel-dégel et des sels fondants devrait cependant considérer ces résultats avec une certaine prudence. Il existe en effet sur le marché une grande diversité de membranes de mûrissement. Or, certains résultats semblent indiquer que l'effet de celles-ci sur la résistance à l'écaillage du béton a tendance à varier en fonction du type de membrane utilisé et des conditions d'application [15]. D'autres recherches sont nécessaires afin de mieux comprendre l'influence du type de mûrissement sur la résistance à l'écaillage du béton.

En industrie, il est pratique courante de mûrir à des températures élevées certaines pièces préfabriquées de béton. Cette technique favorise le développement des résistances mécaniques aux jeunes âges et permet le démoulage rapide des pièces. Certaines études ont cependant indiqué qu'un tel traitement a tendance à modifier de manière considérable la porosité de la pâte de ciment [61-63]. Les pâtes mûries à des températures élevées ont systématiquement une porosité capillaire plus importante et plus grossière. Cette ouverture de la porosité tend généralement à diminuer de manière significative les résistances mécaniques à long terme [64].

Plusieurs études ont démontré que ce type de traitement a également tendance à réduire, de manière significative, la résistance à l'écaillage du béton [15, 65, 66]. L'ouverture de la porosité, associée aux températures de mûrissement élevées, tend en effet à augmenter considérablement la quantité d'eau gelable de la pâte de ciment [67], ce qui contribue à réduire la résistance à l'écaillage. Il semblerait que l'ajout de fumée de silice permette de réduire les effets négatifs de ce type de traitement. Dans le cas des bétons ordinaires, l'ajout de la fumée de silice n'est cependant pas suffisant pour assurer une bonne résistance à l'écaillage. Par contre, dans les BHP, il apparaît que la fumée de silice et la réduction du rapport E/L permettent d'atténuer, de façon considérable, les effets négatifs du mûrissement à des températures élevées. Dans le cas des BHP, l'influence néfaste de ce traitement sur la résistance à l'écaillage serait même minime [66].

À notre connaissance, il n'existe pas d'information concernant l'effet de la température de mûrissement sur la résistance à l'écaillage des bétons fabriqués avec des CV ou des laitiers. Dans le cas des CV, plusieurs études indiquent cependant que l'élévation de la température de la cure tend à activer la réaction d'hydratation des CV et favorise les gains de résistance mécanique aux jeunes âges [72-74]. De plus, contrairement aux bétons fabriqués uniquement avec des ciments Portland, il apparaît que ces gains se maintiennent à long terme. À la lumière de ces résultats, il n'est pas exclu que l'élévation de la température de cure puisse favoriser une augmentation de la résistance à l'écaillage des bétons fabriqués avec certaines CV. Cette hypothèse mériterait d'être vérifiée. Dans l'éventualité où elle s'avérerait véridique, cette caractéristique des CV pourrait être utilisée avec avantage dans le domaine des bétons préfabriqués.

3.2.2 Influence de la durée du mûrissement

De façon générale, il est bien connu que la résistance à l'écaillage des bétons ordinaires s'améliore avec une augmentation de la durée du mûrissement. Passé un certain seuil (entre 14 et 28 jours selon le type de liant utilisé), l'influence de la durée de mûrissement sur le comportement à l'écaillage des bétons ordinaires tend cependant à diminuer de manière considérable. La pâte de ciment s'est alors suffisamment hydratée pour assurer une bonne protection au béton. Certaines études ont même indiqué que la poursuite du mûrissement jusqu'à des âges très avancés peut même quelquefois réduire la résistance à l'écaillage du béton [17, 37]. Ce phénomène est généralement associé au remplissage progressif des bulles d'air à la surface du béton.

Dans la documentation, on retrouve peu de données sur l'influence de la durée de mûrissement sur le comportement à l'écaillage des BHP. Dans l'ensemble, il semblerait que les BHP développent de très bonnes résistances à l'écaillage après seulement quelques jours de mûrissement [21, 22]. Il convient cependant de souligner que les BHP fabriqués avec des CV échappent généralement à cette règle. Comme nous l'avons mentionné précédemment, l'ajout de CV contribue, dans l'ensemble, à réduire de manière significative la résistance à l'écaillage des BHP [17, 26-28]. Dans la plupart des cas, il apparaît même que la poursuite du mûrissement jusqu'à l'âge de 28 jours ne permette pas d'assurer aux BHP avec CV une très bonne résistance à la compression.

3.2.3 Influence des conditions de séchage

Il est bien connu que le séchage a tendance à modifier les propriétés des couches superficielles du béton. Certaines études ont, par exemple, très bien démontré que les mouvements d'eau dus au séchage créent des tensions capillaires importantes dans la pâte de ciment hydraté. Dans le cas de la plupart des bétons, les tensions induites sont souvent suffisantes pour briser les minces parois de C-S-H qui divisent certains pores de la pâte de ciment. Il en résulte une ouverture de la porosité des couches superficielles du béton, c'est-à-dire une augmentation du nombre et de la continuité des pores de grosse dimension [75-77].

En plus de modifier la structure poreuse du béton, le séchage a également pour effet de créer de la microfissuration, celle-ci étant généralement plus intense en peau des ouvrages [78-80]. Lorsque le béton sèche, la pâte de ciment hydraté située près de la surface des structures tend à perdre du volume. Cette modification volumétrique est partiellement empêchée par les granulats et par les couches inférieures moins touchées par le séchage. Il en résulte la formation de microfissures qui viennent s'ajouter à celles déjà induites par les charges de service.

Ces modifications des propriétés en surface peuvent, dans certains cas, affecter la résistance à l'écaillage d'un ouvrage en béton. Dans cette section, nous traitons de l'influence du séchage sur la résistance à l'écaillage des BHP. Les effets de la durée du séchage et de la température de séchage y sont abordés séparément. L'information contenue dans cette section devrait être utile à l'ingénieur civil et au spécialiste en technologie du béton. La majorité des essais d'écaillage normalisés exige, en effet, de sécher les éprouvettes de béton avant de les tester. Cependant, comme on peut le voir dans le Tabl. 1, la durée de la période de séchage tend à varier selon l'essai.

3.2.4   Influence de la durée de séchage

Les données sur l'influence de la durée de séchage que l'on retrouve dans la documentation sont souvent contradictoires. Dans l'ensemble, il semblerait que la résistance à l'écaillage d'un béton ordinaire séché à 20° C et à 50% H.R. pendant une courte période (quelques semaines) puis resaturé soit significativement supérieure à celle d'un béton gardé constamment saturé [81-83]. Selon les résultats de Whiting [30], il apparaît cependant que l'effet bénéfique d'une courte période de séchage a tendance à diminuer avec une diminution du rapport E/L pour devenir pratiquement nul dans le cas des BHP.

Curieusement, d'autres études démontrent également que plusieurs semaines de séchage à 20° C et à 50% H.R. tendent à réduire la résistance à l'écaillage du béton ordinaire [65, 77, 84]. Les résultats de Hammer et Sellevold [14] indiquent, par contre, qu'une augmentation de la durée de séchage a peu d'effet sur le comportement à l'écaillage des BHP. À la lumière des connaissances actuelles, ces apparentes contradictions demeurent difficilement explicables. Globalement, il semble que la durée de séchage ait cependant peu d'effet sur le comportement à l'écaillage des BHP.

3.2.5   Influence de la température de séchage

Au cours des dernières années, plusieurs études ont indiqué qu'une augmentation de la température de séchage a généralement tendance à considérablement diminuer la résistance à l'écaillage des bétons ordinaires [65, 77, 84, 85]. Cet effet est illustré à la Fig. 6. Comme on peut le constater, le séchage à 40° C tend à multiplier par trois la détérioration du béton alors que le séchage à 110° C a un effet encore plus désastreux. Dans l'ensemble, ces résultats s'accordent bien avec les mesures de calorimétrie à basse température qui indiquent que, dans la plupart des cas, la quantité d'eau gelable du béton a tendance à augmenter avec la température de séchage [77].

La plupart des études démontrent également que l'ajout de fumée de silice et la réduction du rapport E/L permettent d'atténuer les effets du séchage à des températures élevées. Ainsi, le séchage à des températures de l'ordre de 40° C à 50° C n'aurait pas d'effet significatif sur la résistance à l'écaillage des BHP [14, 65, 77, 85]. Ces bétons seraient également nettement moins sensibles aux dommages causés par le séchage à 110° C [77].

Dans la documentation, on retrouve très peu d'information sur l'influence de la température de séchage sur le comportement à l'écaillage des bétons avec CV et laitiers. Les seules données disponibles sur le sujet indiquent que l'ajout de certaines CV permet d'atténuer les effets du séchage à 40° C sur le comportement à l'écaillage des bétons de rapport E/L de l'ordre de 0.45 [77]. D'autres études sont cependant nécessaires avant de se prononcer de façon définitive sur l'influence des CV dans ce domaine.

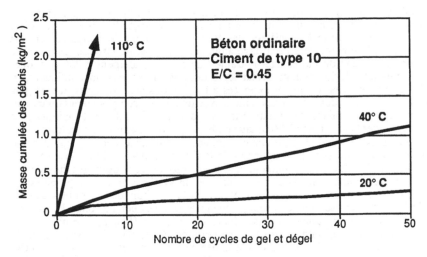

Fig. 6   Effet de la température de séchage sur l'écaillage d'un béton ordinaire

4   Conclusion

Il est relativement difficile de synthétiser le comportement à l'écaillage des BHP et des BTHP. La complexité du problème provient, non seulement de la diversité des paramètres de formulation, mais aussi de la variabilité des caractéristiques physico-chimiques des BHP. À cet ensemble de paramètres viennent aussi s'ajouter tous les facteurs associés au mode de fabrication, au mûrissement et aux conditions d'exposition avant ou pendant l'exposition aux sels fondants.

Pour le moment, la performance relative des BHP et des BTHP exposés aux sels fondants peut difficilement être estimée à partir du comportement de structures existantes. Les structures en BHP exposées aux sels fondants sont encore assez rares et leur âge, en général peu avancé, ne permet pas encore d'en tirer des recommandations pertinentes [68-71]. Il est important de rappeler que la plupart des recommandations qui précèdent sont, par conséquent, fondées sur des comportements observés lors d'essais accélérés d'écaillage en laboratoire. Ces essais (ASTM C 672 et SS 13 72 44) sont certes plus sévères que les conditions d'exposition généralement observées en nature, mais ils ne sont pas nécessairement trop sévères. En effet, on admet généralement que ces deux essais accélérés offrent une alternative relativement simple et rapide pour estimer et comparer la sensibilité à l'écaillage des bétons. Ils permettent notamment d'identifier les combinaisons les plus néfastes. Les recommandations contenues dans certaines normes nationales concernant la formulation des bétons exposés aux sels fondants sont, pour une large part, fondées sur les comportements mesurés à l'aide de ce type d'essai.

On reconnaît maintenant qu'il est possible de fabriquer des bétons qui soient parfaitement en mesure de résister à l'attaque des sels fondants sans la protection d'un réseau de bulles d'air entraîné. En général, c'est la valeur du rapport E/L qui permet de juger de la nécessité ou non d'utiliser un agent entraîneur d'air. Dans le cas des bétons sans ajout, il ne semble pas nécessaire d'utiliser un agent entraîneur d'air si le rapport E/L est de 0.25 ou moins. Un ciment plus performant (haute performance ou haute résistance initiale) ou un ciment avec fumée de silice peuvent permettre d'élever cette limite à 0.30.

Les résultats disponibles jusqu'à présent tendent à démontrer que le remplacement du ciment par une CV ou par un laitier de haut fourneau a généralement pour effet d'augmenter la sensibilité à l'écaillage des BHP. Plus le taux de remplacement est important, plus la sensibilité à l'écaillage augmente. Il apparaît essentiel de toujours prévoir un bon réseau de bulles d'air entraîné (facteur d'espacement < 250 µm) quel que soit le rapport E/L du mélange. L'état actuel des connaissances tend à indiquer que, dans des structures très fortement exposées à l'action des sels fondants, il n'est pas recommandé d'utiliser des CV ou des laitiers avec un taux de remplacement du ciment supérieur à 30%. En raison de la nature très variable des CV, il serait toujours plus prudent de valider la résistance à l'écaillage du béton par des essais en laboratoire.

Les bétons secs ont généralement une plus grande sensibilité à l'écaillage qui découle de leur microstructure plus hétérogène et de la difficulté d'y produire un réseau de bulles d'air entraîné. Il est généralement possible de produire des ouvrages durables en utilisant un agent entraîneur d'air très efficace et un système de malaxage puissant. Dans le cas des bétons sans affaissement et des pavés de béton, de très faibles teneurs en air (2%) permettent souvent de protéger adéquatement ces bétons contre l'écaillage. Si l'ajout d'un agent entraîneur d'air se révèle tout à fait inefficace, il semble que la réduction du rapport E/L en dessous d'une certaine limite (≈ 0.30) permettrait de produire des bétons durables. Les CV et la fumée de silice ont relativement peu d'effet sur la résistance à l'écaillage des bétons sans affaissement et des pavés de béton. Par contre, dans les revêtements routiers en béton compacté au rouleau, la fumée de silice a un effet nettement plus positif et permet souvent de produire des bétons durables sans air entraîné. L'influence des CV sur la résistance à l'écaillage des bétons compactés au rouleau est encore mal comprise et leur emploi devrait être considéré avec un certaine prudence.

Selon les critères exposés au début de cette publication, seuls les BTHP sans ajout ou avec fumée de silice seraient en mesure de résister aux sels fondants sans la protection d'un réseau de bulles d'air. Les normes nationales sur la résistance à l'écaillage peuvent parfois être pénalisantes envers ces BTHP. En effet, pour les conditions d'exposition les plus sévères, elles peuvent imposer l'utilisation d'un fort volume d'air entraîné dans des BTHP possédant déjà un bon comportement à l'écaillage sans cette protection supplémentaire. Les performances mécaniques sont, par conséquent, significativement réduites et elles ne peuvent pas être pleinement mises à profit lors de la conception structurale de l'ouvrage.

La durée de la cure et la durée de la période de séchage qui suit la période de cure ont relativement peu d'influence sur la résistance à l'écaillage des BHP. En effet, certains BHP peuvent développer une très bonne résistance à l'écaillage après de très courtes périodes de mûrissement (< 3 jours). Les BHP avec CV font cependant exception à cette règle. Les BHP résistent généralement beaucoup mieux aux effets néfastes d'un mûrissement accéléré à haute température. L'utilisation simultanée de fumée de silice et d'un rapport E/L faible a souvent pour effet d'annuler presque complètement les effets néfastes des traitements à haute température sur la résistance à l'écaillage du béton.

L'utilisation de BHP sans air entraîné dans des structures soumises à l'action combinée des cycles de gel-dégel et des sels fondants doit se faire en tenant compte d'une très importante règle de prudence. On a déjà mentionné que les exigences minimales concernant le réseau de bulles d'air entraîné diffèrent en fonction du type de destruction par le gel (fissuration interne ou écaillage de la surface). Pour obtenir une structure globalement durable face aux cycles de gel-dégel, il faut donc s'assurer que le choix des caractéristiques du réseau de bulles d'air réponde, non seulement aux exigences concernant la résistance à l'écaillage, mais aussi à celles concernant la résistance à la fissuration interne.

5 Références

1. PHILEO, R. (1987) *Frost susceptibility of high-strength concrete*, Katharine and Bryant Mather International Conference on Concrete Durability, ACI SP-100, J.M. Scanlon Ed., pp. 819-843.
2. MEHTA, P. K. (1991) *Durability of concrete - Fifty years of progress ?*, Second International Conference on Durability of Concrete, ACI SP-126, V.M. Malhotra Ed., pp. 1-32.
3. AÏTCIN, P.-C. (1993) *Durable concrete*, 6th International Conference on Durability of Building Materials and Components, Omya, Japon, **10 p.**
4. MARCHAND, J., SELLEVOLD, E. J., PIGEON, M. (1994) *The deicer salt scaling deterioration of concrete - an overview*, ACI Special Publication SP-145, V.M. Malhotra Ed., pp. 1-46.
5. PIGEON, M., GAGNÉ, R., AÏTCIN, P.-C., LANGLOIS, M. (1992) *La durabilité au gel des bétons à haute performance*, Revue canadienne de génie civil, Vol. 19, N° 6, pp. 975-980.
6. GAGNÉ, R., PIGEON, M., AÏTCIN, P.-C. (1991) *Deicer salt scaling resistance of high-strength concrete made with different cements*, 2e Conférence internationale sur la durabilité du béton, ACI SP-126, V.M. Malhotra Ed., pp. 185-194.
7. GAGNÉ, R., PIGEON, M., AÏTCIN, P.-C. (1990) *Deicer salt scaling resistance of high performance concrete*, ACI SP 122, D. Whiting Ed., pp. 29-44.
8. FOY, C., PIGEON, M., BANTHIA, N. (1988) *Freeze-thaw durability and deicer salt scaling resistance of a 0.25 water-cement ratio concrete*, Cement and Concrete Research, Vol. 18, N° 4, pp. 604-614.
9. BORDELEAU, D., PIGEON, M., BANTHIA, N. (1992) *Latex-modified and normal concretes subjected to freezing and thawing in the presence of deicer salt solution*, ACI Materials Journal, Vol. 89, N° 6, pp. 547-553.
10. UCHIKAWA, H (1994) *Durability of high strength concrete with superior workability estimated from the composition and Structure*, **Accepté pour publication dans les comptes rendus de P.K. Metha Symposium on Concrete Durability, Nice, France, 30 p.**
11. PIGEON, M. (1987) *La durabilité au gel du béton* , Materials and Structures/ Matériaux et constructions, Vol. 22, pp. 3-14.
12. ASTM (1989) *Annual book of ASTM standards, Part 13: Cement, lime, celling and walls (including manual of cement testing)*, American Society for Testing and Materials, Philadelphia.
13. SIS (1992) *Concrete testing - Hardened concrete - Frost resistance*, Standard-iserings-kommissionen i Sverige, Swedish Standard SS 13 72 44.
14. HAMMER, T. A., SELLEVOLD E. J. (1990) *Frost resistance of high-strength concrete*, Second International Conference on High-Strength Concrete, ACI SP-121, W.T. Hester Ed., pp. 457-487.
15. LANGLOIS M., BEAUPRÉ, D., PIGEON, M., FOY, C. (1989) *The influence of curing on the salt scaling resistance of concrete with and without silica fume*, ACI SP-114, V.M. Malhotra Ed., pp. 971-990.
16. RICKNE, S., NYQVIST, H. (1989) *Salt frost resistant concrete exposed to deicing salts* , 9th European Ready Mixed Concrete Organization Congress, ERMCO'89, Stavenger, Norvège, juin, pp. 56-91.
17. FAGERLUND, G. (1986) *Effect of air-entraining and other admixtures on the salt-scaling resistance of concrete*, Chalmers International Seminar on some Aspects of Admixture and Industrial By-Products on the Durability of Concrete, Göteburg, Suède, 33 p.
18. BILODEAU A., CARETTE, G. G. (1989) *Resistance of condensed silica fume concrete to the combined action of freezing and thawing cycling and deicing salts*, ACI SP-114, V.M. Malhotra Ed., pp. 945-969.

19. PETERSSON, P.E. (1986) *The influence of silica fume on the salt frost resistance of concrete*, Rapport technique SP-RAPP 1986:32, Swedish National Testing Institute, Division of Building Technology, 18 p.

20. PIGEON, M., GAGNÉ, R., MARCHAND, M., BOISVERT, J., HORNAIN H. (1991) *La durabilité au gel du béton à hautes performances*, Séminaire sur les bétons à hautes performances, École Normale Supérieure de Cachan, France, 15 p.

21. GAGNÉ, R., PIGEON, M., AÏTCIN, P.-C. (1990) *Deicer salt scaling resistance of high-performance concrete*, ACI SP-122, David Whiting Ed., pp. 29-44.

22. GAGNÉ, R. (1992) *La durabilité au gel des bétons à hautes performances*, Thèse de doctorat, Université Laval, Département de génie civil, Sainte-Foy (Québec), Canada, 433 p.

23. PIGEON, M., PERRATON, D., PLEAU, R. (1987) *Scaling test of silica fume concrete and the critical spacing factor concept*, Katharine and Bryant Mather International Conference on Concrete Durability, ACI SP-100, J. M. Scanlon Ed., pp. 1155-1182.

24. VIRTANEN, J. (1990) *Field study on the effects of additions on the salt scaling resistance of concrete*, Nordic Concrete Research, Vol. 9, pp. 197-212.

25. AÏTCIN, P.-C., PIGEON, M. (1986) *Performance of condensed silica fume concrete used in pavements and sidewalks*, Durability of Building Materials, Vol. 4, N° 3, pp. 353-368.

26. BILODEAU, A., CARETTE, G. G., MALHOTRA, V. M., LANGLEY, W. S. (1991) *Influence of curing and drying on salt scaling resistance of fly ash concrete*, ACI SP-126, V. M. Malhotra Ed., pp. 210-228.

27. BILODEAU A., MALHOTRA M. (1992) *Concrete incorporating high-volumes of ASTM class F fly ash: mechanical properties and resistance to deicing salt scaling and to Chloride-ion penetration* , ACI SP-132, V. M. Malhotra Ed., pp. 319-330.

28. REID, É., PIGEON, M., PLEAU, R. (1993) *La durabilité au gel des bétons à haute performance contenant des cendres volantes de classe F*, Demi-journée ouverte du Réseau de centres d'excellence sur les bétons à haute performance, Québec, juin, pp. 19-34.

29. MALHOTRA, V. M. (1992) *CANMET investigation dealing with high-volume fly ash concrete*, Advances in Concrete Technology, V. M Malhotra Ed., CANMET, Energy mines and resources Canada, pp. 433-470.

30. WHITING, D. (1987) *Durability of high-strength concrete*, ACI SP-100, J. M. Scanlon Ed., pp. 169-183.

31. BARROW, R. S., HADCHITI, K. M., CARRASQUILLO, P.M., CARRASQUILLO, R.L. (1989) *Temperature rise and durability of concrete containing fly ash*, ACI SP-114, V.M. Malhotra Ed., pp. 331-347.

32. KLEIGER, P., GEBLER, S. (1987) *Fly ash and concrete durability*, ACI SP-100, J. M. Scanlon Ed., pp. 1043-1069.

33. JOHNSTON, C. (1987) *Effects of microsilica and class C fly ash on resistance of concrete to rapid freezing and thawing and scaling in the presence of deicing agents*, ACI SP-100, J. M. Scanlon Ed., pp. 1183-1204.

34. BRUNO, V., CHABANET, M., AMBROISE, J., PERA, J. (1991) *Bétons à très hautes performances (B.T.H.P.) : microfissuration et durabilité*, 2ᵉ Conférence internationale CANMET/ACI sur la durabilité du béton, Montréal, Canada, Supplementary papers, pp. 175-194.

35. BYFORS, G., KLINGSTEDT, G., LEHTONEN, V., PYY, H., ROMBEN, L. (1989) *Durability of concrete made with alkali activated slag*, ACI SP-114, V.M. Malhotra Ed., pp. 1429-1455.

36. FAGERLUND, G. (1982) *The influence of slag cement on the frost resistance of the hardened concrete*, CBI, Research 1:82, Swedish Cement and Concrete Institute, Stockholm.

37. GUNTER, M., BIER, TH., HILSDORF, H. (1987) *Effect of curing and type of cement on the resistance of concrete to freezing in deicing salt solution*, ACI SP-100, J. M. Scanlon Ed., pp. 877-899.
38. BAALBAKI, M. (1989) *Ciment spécial pour bétons à haute performance*, Mémoire de maîtrise, Université de Sherbrooke, Département de génie civil, Sherbrooke, Canada, 130 p.
39. NISCHER, P. (1974) *Production and control of air-entrained concrete for concrete blocks*, Concrete Precasting Plant and Technology, Vol. 34, N° 12, pp. 34-38.
40. WHITING, D. (1985) *Air contents and air-void characteristics of low-slump dense concretes*, ACI Materials Journal, Vol. 82, pp. 716-723.
41. JACOBSEN, S., FARSTAD, T., GRAN, H.C., SELLEVOLD, E.J. (1992) *Frost/salt scaling of no-slump concrete: Effect of strength, air-entraining agent and condensed silica fume*, Nordic Concrete Research, Vol. 11, 15 p.
42. BOISVERT, J., MARCHAND, J., PIGEON, M., ISABELLE, H.L. (1992) *Durabilité au gel-dégel et résistance à l'écaillage des pavés de béton*, Revue canadienne de génie civil, Vol. 19, pp. 1017-1024.
43. MARCHAND, J., TREMBLAY, S., BOISVERT, L., MALTAIS, J., PIGEON, M. (1994) *Air entrainment in dry concretes*, Atelier international sur les bétons compactés au rouleau, Département de génie civil, Université Laval, pp. 83-100.
44. GOMEZ-DOMINGUEZ, J. (1988) *Roller compacted concrete for highway applications*, Ph/. D. Thesis, School of Civil Engineering, Purdue University, 236 p.
45. HORRIGMOE, G., BROX RINDAL, D. (1990) *High strength roller compacted concrete*, 6e Symposium International des Routes en Béton, CEMBUREAU, Belgique, pp. 51-60.
46. RAGAN, S.A. (1991) *The use of air entrainment to ensure the frost resistance of roller-compacted concrete pavements*, ACI SP-126, pp. 115-130.
47. MARCHAND, J., PIGEON, M., ISABELLE, H.L., BOISVERT, J. (1990) *Freeze-thaw durability and deicer salt scaling resistance of roller-compacted concrete pavements*, ACI SP-122, D. Whiting Ed., pp. 217-236.
48. MARCHAND, J., BOISVERT, J., PIGEON, M., ISABELLE, H.L. (1991) *Deicer salt scaling resistance of roller-compacted concrete pavements*, ACI SP-126, V.M. Malhotra Ed., pp. 131-153.
49. MARCHAND, J., PIGEON, M., BOISVERT, J., ISABELLE, H.L., HOUDUSSE, O. (1992) *Deicer salt scaling resistance of roller-compacted concrete pavements containing fly ash and silica fume*, ACI SP-132, V.M. Malhotra Ed., pp. 151-178.
50. PIGEON, M., MARCHAND, J. (1993) *The frost durability of dry concrete products*, Rapport de recherche GCS-93-06, Département de Génie Civil, Université Laval, 196 p.
51. MARCHAND, J., HORNAIN, H., DIAMOND, S., PIGEON, M., GUIRAUD, H. (1996) *The microstructure of dry concrete products*, Cement and Concrete Research, Vol. 26, N° 3, pp. 427-438.
52. CLARK, A.J. (1980) *Freeze-thaw durability upon concrete paving block specimens*, First International Conference on Concrete Block Paving, University of New-Castle-Upon-Tyne, Angleterre, pp. 106-112.
53. SIEBEL, E., NECK, U. (1990) *Durability of paving setts and flagstones*, Concrete Precasting Plant and Technology, Vol. 56, N° 8, pp. 34-38.
54. ANDERSSON, R. (1986) *Pavements of roller-compacted concrete - physical properties*, Nordic Concrete Research, Vol. 5, pp. 7-17.
55. BROX RINDAL, D., HORRIGMOE, G. (1993) *High quality roller compacted concrete pavements*, Third International Conference on the Utilisation of High Strength Concrete, Lillehammer, Norvège, pp. 913-921.

56.   BANTHIA, N., PIGEON, M., MARCHAND, J., BOISVERT, J. (1992) *Water permeability of roller-compacted concrete pavements*, ASCE Materials Journal, Vol. 4, N° 1, pp. 27-40.
57.   ANDERSSON, R. (1987) *Swedish experiences with RCC*, Concrete International, Vol. 9, N° 2, pp. 18-24.
58.   POWERS, T.C. (1961) *Some physical aspects of the hydration of Portland cement*, Journal of the PCA Research and Development Laboratories, Vol. 3, N° 1, pp. 47-56.
59.   PATEL, R.G., KILLOH, D.C., PARROTT, L.J., GUTTERIDGE, W.A. (1988) *Influence of curing at different relative humidities upon compound reactions and porosity in Portland cement paste*, Materials and Structures/Matériaux et Constructions, Vol. 21, pp. 192-197.
60.   AFRANI, I., ROGERS, C. (1993) *The effect of different cementing materials and curing regimes on the scaling resistance of concrete*, Troisième colloque canadien sur les ciments et bétons, 3-4 août, Ottawa, Canada, 28 p.
61.   RADJY, F., RICHARDS, C. (1973) *Effect of curing and heat treatment history on the dynamic mechanical response and the pore structure of hardened cement paste*, Cement and Concrete Research, Vol. 3, N° 1, pp. 7-21.
62.   BRAY, W.H., SELLEVOLD, E.J. (1973) *Water sorption properties of hardened cement paste cured and stored at elevated temperatures*, Cement and Concrete Research, Vol. 3, N° 6, pp. 723-728.
63.   SELLEVOLD, E.J. (1974) *Mercury porosimetry of hardened cement paste cured or stored at 97° C*, Cement and Concrete Research, Vol. 4, N° 3, pp. 399-404.
64.   LINDGÅRD, J., SELLEVOLD, E.J. (1993) *Is high-strength concrete more robust against elevated curing temperatures?*, Third International Conference on the Utilization of High Strength Concrete, Lillehammer, Norvège, pp. 810-820.
65.   SELLEVOLD, E.J., FARSTAD, T. (1991) *Frost/salt testing of concrete: effect of test parameters and concrete moisture history*, Nordic Concrete Research, Vol. 10, pp. 121-138.
66.   JACOBSEN, S., SELLEVOLD, E.J. (1992) *Frost/salt scaling of concrete: effect of curing temperature and condensed silica fume on normal and high-strength concrete*, Fourth ACI/CANMET International Conference on Fly Ash, Silica Fume, Slag and Natural Pozzolans in Concrete. Supplementary Papers, pp. 369-384.
67.   SELLEVOLD, E.J., BAGER, D.H. (1985) *Some implications of calorimetric ice formation results for frost resistance testing of cement products*, Technical Report 86/80, The Technical University of Denmark, Building Materials Laboratory, 28 p.
68.   HOFF, G. (1993) *Utilization of high-strength concrete in North America*, Proceedings of the Symposium on the Utilization of High-Strength Concrete, Lillehammer, Norvège, juin, pp. 28-36.
69.   KÖNIG, H., BERGNER, H., GRIMM, R., SIMSCH, G. (1993) *Utilization of high-strength concrete in Europe*, Proceedings of the Symposium on the Utilization of High-Strength Concrete, Lillehammer, Norvège, juin, pp. 45-56.
70.   HOLAND, I. (1993) *High-strength concrete in Norway. utilization and research*, Proceedings of the Symposium on the Utilization of High-Strength Concrete, Lillehammer, Norvège, juin, pp. 68-82.
71.   MALIER, Y. (1992) *High-performance concrete: From material to structure - Introduction*, E & FN Spon, Éd. Yves Malier, 542 p.
72.   FELDMAN, R.F. (1983) *Significance of porosity measurements on blended cement performance*, ACI SP-79, V.M. Malhotra Ed., pp. 415-433.
73.   BERRY, E.E., MALHOTRA, V.M. (1987) *Utilisation des cendres volantes dans la préparation du béton*, Matériaux complémentaires en cimentation pour le béton, CANMET, V.M. Malhotra Ed., pp.39-181.

74. RAVINA, D. (1992) *Temperature and curing effects on concrete with high volumes of fly ash*, CBI/CANMET International Symposium on the Use of Fly Ash, Silica Fume, Slag and other By-Products in Concrete and Construction Materials, 2-4 novembre, Milwaukee, États-Unis, 11 p.

75. SELLEVOLD, E.J., RADJY, F. (1976), *Drying and resaturation effects on internal friction in hardened cement pastes*, Journal of the American Ceramic Society, Vol. 59, N° 5-6, pp. 256-258.

76. FELDMAN, R.F. (1988) *Effect of pre-drying on rate of water replacement from cement paste by Propan-2-ol*, Il Cemento, N° 3, pp. 193-201.

77. MARCHAND, J. (1993) *Contribution à l'étude de la détérioration par écaillage du béton en présence de sels fondants*, Thèse de Doctorat, École Nationale des Ponts et Chaussées, Paris, France, 316 p.

78. BAZANT, Z.P., RAFTSHOL, W.J. (1982) *Effect of cracking in drying and shrinkage specimens*, Cement and Concrete Research, Vol. 12, N° 1, pp. 209-226.

79. HWANG, C.L., YOUNG, J.F. (1984) *Drying shrinkage of Portland cement pastes — Part I: Microcracking during drying*, Cement and Concrete Research, Vol. 14, pp. 585-594.

80. ACKER, P. (1992) *Retraits et fissurations du béton*, Association Française pour la Construction, 42 p.

81. VERBECK, G.J., KLIEGER, P. (1957) *Studies of "salt" scaling of concrete*, Highway Research Board Bulletin, N° 150, pp. 1-13.

82. WONG, A.Y.C., ANDERSON, C.L., HILSDORF, H.K. (1973) *The effect of drying on the freeze-thaw durability of concrete*, Engineering Experiment Station Bulletin 506, University of Illinois, 41 p.

83. PETERSSON, P.E. (1992) *Scaling resistance of concrete — Influence of curing conditions*, Réunion du Groupe de Travail TC 117 de la RILEM, Dübendorf, octobre.

84. LAROCHE, M.-C., MARCHAND, J., PIGEON, M. (1993) *Fiabilité de l'essai ASTM C 672 : Comparaison entre la résistance à l'écaillage du béton mesurée en laboratoire et son comportement en service*, Atelier international sur la résistance des bétons aux cycles de gel-dégel en présence de sels fondants, Comité RILEM TC 117, Québec, 30-31 août, 17 p.

85. SORENSEN, E.G. (1983) *Freezing and thawing resistance of condensed silica fume (microsilica) concrete exposed to deicing chemicals*, ACI SP-79, V.M. Malhotra Ed., pp. 709-718.

6

Frost/salt scaling and ice formation of concrete: effect of curing temperature and silica fume on normal and high strength concrete

S. JACOBSEN
The Norwegian Building Research Institute, Oslo, Norway

E.J. SELLEVOLD
The Norwegian Inst. of Technology, Trondheim, Norway

Abstract
The present experiments were carried out to determine the consequences of the temperature and moisture histories for the ice formation and frost/salt scaling resistance of concrete. Non-airentrained concrete mixes with W/(C+S) = 0.30 and 0.45, both with and without 8 % silica fume, were tested. The results show that elevated curing temperatures (60 °C) drastically increase the scaling for the two W/(C+S) = 0.45 mixes, regardless of the early temperature history. (The period at 20 °C before heating to 60 °C was varied from 0 to 7 hours). 40 °C gives a smaller increase in scaling. For the W/(C+S) = 0.30 mixes the effects of elevated curing temperatures are much less pronounced, and the level of scaling much lower even after a dry/rewet treatment before testing. Silica fume reduces the scaling for both W/(C+S) - ratios and all curing procedures. Low temperature calorimetry has shown that ice formation down to - 20 °C is very small for concretes with W/(C+S) = 0.30 and S/(C+S) = 0.08 both for 20°C and 60 °C curing. Concrete with w/c = 0.30 without SF showed somewhat more ice-formation both after 60 °C and 20 °C curing. The results showed that high strength concrete with silica fume can be frost/salt resistant without air entrainment, and the results support earlier findings, that there is at least a qualitative correlation between ice formation and frost/salt scaling of concrete.
Keywords: concrete, durability, frost/salt scaling, heat curing, silica fume, ice formation

1 Introduction

Measurements of ice formation in water saturated hardened cement paste (HCP) has shown that both drying/rewetting [1,2] and curing at elevated temperatures [3] lead to substantial increases in ice formation as measured in a low-temperature scanning calorimeter. The test [3] was performed on HCP with W/(C+S) = 0.40 and 0 and 8 %

Freeze-Thaw Durability of Concrete. Edited by J. Marchand, M. Pigeon and M. Setzer.
Published in 1997 by E & FN Spon, 2–6 Boundary Row, London SE1 8HN, UK.
ISBN 0 419 20000 2.

silica fume (SF). The ice formation increased with increasing curing temperature and decreased with the addition of SF.

In the case of drying/rewetting treatment the increased ice formation has been shown to result in decreased frost resistance [4]. Figure 1 shows the effects of drying/resaturation on ice formation and frost/salt scaling, and elevated curing temperatures on ice formation. The consequences of elevated curing temperatures for frost/salt scaling resistance has not been investigated systematically so far, and this was the purpose of the experiments described in this paper. The tests included both normal quality concrete (W/C = 0.45) and high strength concrete (W/C = 0.30) both with and without 8 % of the cement replaced with SF on a weight basis. In the Nordic countries frost/salt scaling tests are considered to be most relevant since problems with frost deterioration are connected with the use of deicing salts.

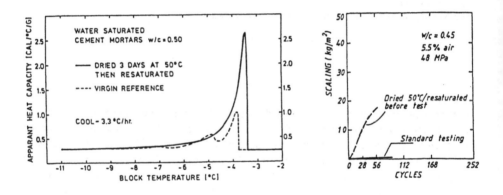

Fig. 1 a and b.   Effect of drying/resaturation on ice formation in mortar [1].
                   Effect of drying resaturation on salt/frost scaling of concrete [4].

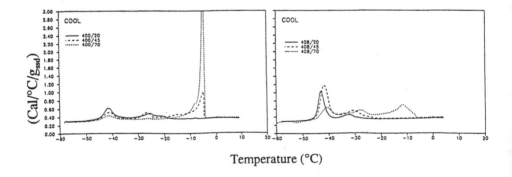

Fig. 1 c.   Effect of heatcuring on ice formation in HCP [3].

2 Concrete mixtures

The concrete mixtures used are given in Table 1. The W/(C+S) = 0.45 mixes were made with a Norwegian RP 38 rapid type portland cement, and the W/(C+S) = 0.30 mixes were made with a Norwegian P 30 - 4A special high strength cement for off-shore structures. In addition one extra mixture with W/(C+S) = 0.30 and 8 % SF with a Norwegian P 30 normal portland cement was tested. The SF used in the W/(C+S) = 0.30 mixtures is from Fiskaa Norway, and the SF used in the W/(C+S) = 0.45 is from Hafslund Norway. Characteristics of cement and silica fume are given in Table 2. The W/(C+S) = 0.45 mixes are from the same batches as the corresponding mixes presented in [5]. Details of the W/(C+S) = 0.30 mixes are also given elsewhere, [6].

Table 1. Concrete mixtures

Mix no.	045 - 00	045 - 08	030 - 00	030 - 08	030 - 08P
W/(C+S)	0.45	0.45	0.30	0.30	0.30
S/(C+S)	0	0.08	0	0.08	0.08
Type of cement	RP 38	RP 38	P 30-4A	P 30-4A	P 30
Cement	380	350	648	585.5	585.5
SF	-	30	-	51	51
Water (total)	171	171	194.5	191	191
Aggregate:					
0 - 0.125	-	-	113	113	113
0.125 - 8	-	-	1495	1495	1495
0 - 8	1023	1023	-	-	-
8 - 16	817	817	-	-	-
Admixtures					
(40% solid)					
Lignosulph.	2.3	2.3	-	-	-
Melamin	5.3	5.3	-	-	-
Naphtalen	-	-	9.1	9.9	14.3

All concrete mixtures were made without air-entraining admixture. The following abbreviations are used for the different concrete mixtures:

045 - 00	W/(C+S) = 0.45, S/(C+S) = 0	(RP 38 cement)
045 - 08	W/(C+S) = 0.45, S/(C+S) = 0.08	(RP 38 cement)
030 - 00	W/(C+S) = 0.30, S/(C+S) = 0	(P 30-4A cement)
030 - 08	W/(C+S) = 0.30, S/(C+S) = 0.08	(P 30-4A cement)
030 - 08 P	W/(C+S) = 0.30, S/(C+S)= 0.08	(P 30 cement)

Table 2. Composition of cement and SF

	Cement			SF	
	RP 38	P 30-4A	P 30	Hafslund	Fiskaa
Fineness (m^2/kg)	509	420	383	-	-
Setting time min.					
- initial	120	183	97	-	-
- final	164	230	138	-	-
Comp. strength (MPa)					
- 1 day	32.4	21.0	20.1	-	-
- 3 days	45.1	33.5	33.1	-	-
- 7 days	50.8	43.0	40.1	-	-
-28 days	58.9	63.6	49.3	-	-
Chem.composition (%)					
- SiO$_2$	20.6	22.1	20.7	88.2	95
- Al$_2$O$_3$	4.9	4.2	5.0	1.0	-
- Fe$_2$O$_3$	3.0	3.4	2.9	2.4	-
- CaO	63.5	64.1	63.7	0.6	-
- MgO	1.7	1.4	1.7	2.0	-
- SO$_3$	3.4	2.7	2.9	-	-
- K$_2$O	0.8	0.6	0.8	1.9	-
- Na$_2$O	0.7	0.2	0.7	0.8	-
- L.O.I.	1.2	0.9	1.1	-	
Mineral Composition					
- C$_3$S	55.9	52.4	56.7	-	-
- C$_2$S	16.8	23.7	16.3	-	-
- C$_3$A	7.9	5.5	8.3	-	-
- C$_4$AF	9.0	10.3	9.0	-	-

3 Experimental procedures

3.1 Curing

Cubes 100 mm*100 mm*100 mm and cylinders of diameter 100 mm and height 200 mm were cast and heat cured by storing the specimens in three different water tanks kept at 20, 40 and 60 °C. Specimens from each concrete mixture were given different temperature - time curing procedures after casting, by moving the specimens from one water container to another at different intervals. The following heat cures were used:

Cure 1:   At 20 °C for 28 days. The specimens were demoulded after 24 hours and then stored in water.

Cure 2:   Directly to 40 ° C 1 hour after casting, then after one hour demoulded and placed at 60 °C for 1 week, thereafter cooled slowly to 20 °C.

Cure 3:   At 20 ° for 7 hours, then at 40 °C for 1 hour, then at 60 °C for 1 week, thereafter cooled to 20 °C. The specimens were carefully demoulded after 7 hours.

Cure 4:   Directly to 40 °C for 1 week after casting, thereafter cooled to 20 °C. The specimens were demoulded after one week.

Cure 5:   At 20 °C for 7 hours, then at 40 °C for one week, thereafter
           cooled to 20 °C. The specimens were carefully demoulded after
           7 hours.

The selection of curing histories was motivated by investigations reported by Lindgård
and Sellevold, [6]. In Table 3 the different combinations of concrete mixtures and
temperature - time curing procedures that were tested are shown.

Table 3. Concrete mixtures and curing methods used in the experiments

			Concrete mix		
	045 - 00	045 - 08	030 - 00	030 - 08	030 - 08 P
Cure 1	x	x	x	x	x
Cure 2	x	x	x	x	x
Cure 3	x	x	x	x	x
Cure 4	x	x			
Cure 5	x	x			

3.2 Frost/salt scaling tests

Frost/salt scaling tests were performed according to Swedish Standard SS 13 72 44
("Borås-method") on specimens cut from the cylinders. After 28 days of curing the
specimens were stored in water for approximately 2 months. Then two cylinders from
each combination of concrete mixtures and curing methods were cut in 50 mm thick
slices. Six parallel specimens, all with sawn surfaces, were tested for each mixture -
cure combination. After sawing the cylinders the W/(C+S) = 0.45 and W/(C+S) = 0.30
mixtures were given different pretreatments before the frost/salt scaling test started.

The 045 - 00 and 045 - 08 series were stored in 50 % relative humidity and 23 °C for
11 days, then prepared as shown in Figure 2 and stored for 3 days with fresh water on
the surface of the specimens. Then the fresh water was removed and a 3 mm layer of 3
% NaCl solution was poured on the surface. The freeze/thaw test started immediately
afterwards according to the temperature - time cycle shown in Figure 2. Scaled off
concrete was measured after 7, 14, 28, 42 and 56 cycles and calculated per unit area.
This is the procedure described by SS 13 72 44.

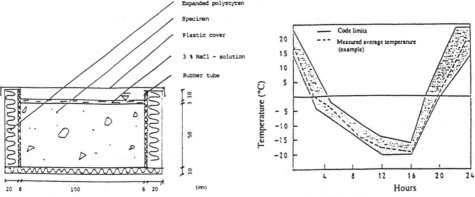

Fig. 2. Specimen preparation and freeze/thaw cycle for SS 13 72 44 frost/salt test

The 030 - 00, 030 - 08 and 030 - 08 P specimens were dried for 14 days in a ventilated oven at 50 °C, cooled, resaturated in 20 °C fresh water for 14 days. The specimens were then prepared as shown in Figures 2 and a 3 mm layer of 3 % NaCl solution was poured on the surface. The freeze/thaw testing started immediately afterwards according to the temperature-time cycle shown in Figure 2.

The drying/rewetting treatment is not described in the standard test method, but as already mentioned, has been found to be a very severe preconditioning with respect to salt/frost scaling. Furthermore, the specimens were run up to a total of 280 freeze/thaw cycles, since it has been found that SF in high strength concrete can lead to high scaling beyond the 56 cycles prescribed in the Borås method [7]. This procedure was chosen in order to provide a very severe test of the salt/frost-scaling resistance of high strength concrete without airentrainment.

3.3 Calorimetric measurements
Low temperature calorimetric measurements were performed on 030 - 00 and 030 - 08, both cured at 20 °C and 60 °C (cure 1 and cure 3) after drying/resaturation treatment. The results were used to calculate the amounts of ice formed in different temperature intervals. The experiments were performed at the Danish Institute of Technology Lyngby, Laboratory of Building Materials. The results are presented as calculated amounts of ice, and plots of apparent heat capacity vs. temperature. The area under the curves (over baseline) are proportional to the amounts of ice formed to a given temperature.

3.4 Compressive strength
Concrete compressive strength was tested on three parallel 100 mm*100 mm*100 mm cubes for each concrete mixture and curing method.

3.5 Air content of hardened concrete
Air content of hardened concrete was measured according to the PF-method, [8]. The testing procedure consists of measuring the weight of a concrete specimen in different moisture conditions, measuring the volume and then calculating porosities and densities from these data:

Weight A: Dried to constant weight (105 °C)
Weight B: Water saturated by capillary suction immersed in water
Weight C: Pressure saturated (10 MPa water pressure)
Volume V: Weighing in water

The following porosities are calculated:

Suction porosity $= (B - A)/ V$ (%)
Air content $= (C - B)/ V$ (%)

The suction porosity determined this way is a function of the volume fraction of cement paste, the W/C - ratio and the degree of hydration assuming that no water is absorbed by the aggregate. The air content is represented by pores that are not filled after immersion in water (capillary suction), but are filled after application of 10 MPa water pressure. In addition different densities can be calculated from the data. In [8] it has been documented that air content measured with the PF-method and air content measured optically on plane sections correlate very well. Air content was measured on four parallel specimens from each combination of concrete mixture and curing method. The specimens were cut from top and bottom of the same cylinders as the frost/salt testing specimens were cut from.

4 Results and discussion

The mass of scaled-off concrete is plotted as cumulative scaling vs. number of cycles in Figures 3, 4, 5, 6 and 7, and shown in Table 4. In Table 5 the 28-day compressive strength is shown. Air-content of the hardened concrete is shown in Table 6.

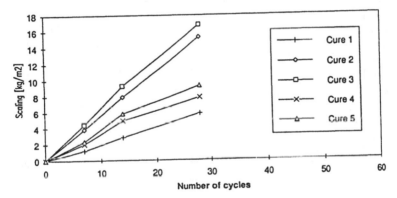

Fig. 3. 045 - 00 Concrete: Number of Cycles vs. Scaling

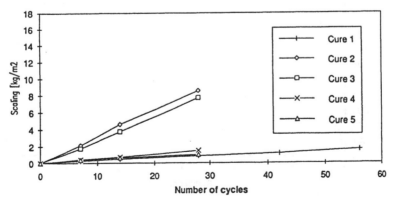

Fig. 4. 045 - 08 Concrete: Number of Cycles vs. Scaling

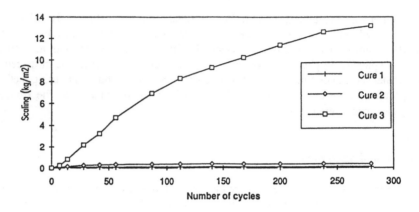

Fig. 5. 030 - 00 Concrete: Number of Cycles vs. Scaling

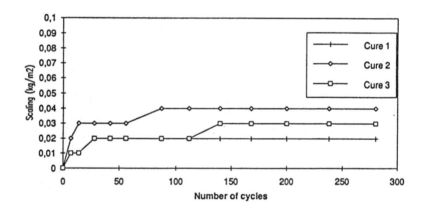

Fig. 6. 030 - 08 Concrete: Number of Cycles vs. Scaling

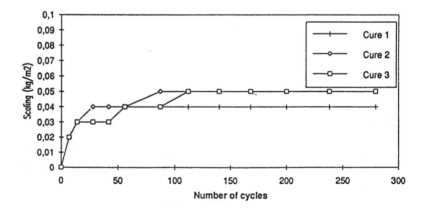

Fig. 7. 030 - 08 P Concrete: Number of Cycles vs. Scaling

Table 4. Scaled concrete in SS 13 72 44 frost/salt test (6 specimens for each series)

Concrete mix		7	s1)	14	s	28	s	56	s	140	s	280	s
045-00	Cure 1	1.21	0.41	2.87	0.66	5.86	0.85	2)		-		-	
	Cure 2	3.81	0.49	7.92	1.21	15.3	1.06	2)		-		-	
	Cure 3	4.42	0.58	9.31	1.36	16.8	1.86	2)		-		-	
	Cure 4	2.02	0.26	4.94	0.54	7.86	1.52	2)		-		-	
	Cure 5	2.33	0.34	5.83	0.63	9.34	0.68	2)		-		-	
045-08	Cure 1	0.26	0.05	0.48	0.04	0.87	0.15	1.67	0.21	-		-	
	Cure 2	2.15	0.25	4.62	0.50	8.57	1.10	2)		-		-	
	Cure 3	1.78	0.60	3.80	0.85	7.72	0.59	2)		-		-	
	Cure 4	0.45	0.09	0.75	0.33	1.51	0.62	2)		-		-	
	Cure 5	0.39	0.13	0.65	0.27	1.03	0.40	2)		-		-	
030-00	Cure 1	0.03	0.01	0.06	0.03	0.08	0.02	0.09	0.00	0.10	0.04	0.10	0.05
	Cure 2	0.06	0.03	0.14	0.08	0.26	0.14	0.34	0.05	0.38	0.34	0.38	0.34
	Cure 3	0.25	0.13	0.82	0.50	2.18	0.88	4.68	0.88	9.26	4.88	13.2	7.99
030-08	Cure 1	0.01	0.00	0.01	0.01	0.02	0.00	0.02	0.00	0.02	0.01	0.02	0.01
	Cure 2	0.02	0.01	0.03	0.02	0.03	0.00	0.03	0.00	0.04	0.01	0.04	0.01
	Cure 3	0.01	0.00	0.01	0.01	0.02	0.00	0.02	0.00	0.03	0.02	0.03	0.02
030-08 P	Cure 1	0.02	0.00	0.03	0.01	0.04	0.00	0.04	0.00	0.04	0.01	0.04	0.01
	Cure 2	0.02	0.00	0.03	0.01	0.04	0.00	0.04	0.00	0.05	0.01	0.05	0.01
	Cure 3	0.02	0.01	0.03	0.01	0.03	0.00	0.04	0.01	0.05	0.02	0.05	0.02

Scaled off-mass (kg/m^2)

1): Standard deviation
2): Testing terminated due to high scaling and leakage of salt solution

Table 5. 28-day compressive strength

Concrete mix	Cure 1 fc	s 1)	Cure 2 fc	s	Cure 3 fc	s	Cure 4 fc	s	Cure 5 fc	s
045 - 00	60.2	1.9	52.4	1.7	53.0	1.3	52.0	0.5	55.5	0.5
045 - 08	63.0	1.7	57.3	1.9	61.2	1.8	60.2	2.1	62.1	1.8
030 - 00	98.5	1.3	-	-	107.7	2.2	-	-	-	-
030 - 08	109.8	0.7	-	-	111.8	0.8	-	-	-	-
030 - 08P	86.2	1.7	-	-	90.7	2.5	-	-	-	-

28 day compressive strength (MPa)

1): Standard deviation

Table 6. Air content measured on hardened concrete with the PF-method [8]

Concrete mix	Cure 1 A	s	Cure 2 A	s	Cure 3 A	s	Cure 4 A	s	Cure 5 A	s
045 - 00	1.5	0.1	1.5	0.2	1.5	0.1	1.0	0.1	1.0	0.1
045 - 08	1.5	0.2	2.0	0.3	1.7	0.3	1.4	0.1	1.2	0.1
030 - 00	1.9	0.0	-	-	1.8	0.4	-	-	-	-
030 - 08	2.3	0.2	-	-	2.4	0.3	-	-	-	-
030 - 08 P	2.4	0.2	-	-	2.2	0.2	-	-	-	-

Air content (volume %)

For the 045 - 00 and 045 - 08 mixtures, Cure 2 and Cure 3 (60 °C) lead to increased scaling compared to Cure 1 (20 °C). Cure 4 and Cure 5 (40 °C) give a small increase in scaling compared to Cure 1 (20 °C), somewhat more for the mixture without SF, Figures 3 and 4. The two different procedures of increasing the temperature give no significant differences in scaling. This is somewhat surprising since the short delay period before heating has been shown to decrease the strength more than if a 6 hours delay period is used before increasing the temperature, [5].

The substitution of 8 % of the cement with SF reduces the scaling for all curing procedures.

For the 030 - 00 series Cure 1 (20 °C) (Fig. 5) gives very low scaling, Cure 2 (60 °C, fast heating) gives a little higher scaling, whereas Cure 3 (60 °C, slow heating) results in a very high scaling. The scattering of results between six single specimens for the 03 - 00 Cure 3 specimens is higher than for any other series in this investigation. For the 030 - 08 and 030 - 08 P the scaling is very low for all curing procedures. The differences in scaling between the different curing procedures and the two types of cement lies within the standard deviation for each series of six parallel specimens.

The results show that substitution of 8 % of the cement with SF results in an improved salt/frost scaling resistance both after drying/rewetting treatment and curing at high temperatures. The results also show that the 030 - 00 concrete mixes cured at 20 °C and 60 °C, fast heating, have a very good salt/frost resistance even after a dry/rewet treatment. The higher scaling after 60°C, slow heating treatment, appears somewhat anomalous and we have no explanation for this result.

The air content for the 045 - 00 and 045 - 08 series varies from 1.0 to 2.0 %. For the 030 - 00, 030 - 08 and 030 - 08 P series the air content varies between 1.8 and 2.4 %. The variation in air content between four parallel samples for each combination of concrete mixture and cure is low.

For the W/(C+S) = 0.45 mixtures the 28 day compressive strength is reduced for all heat cures compared to curing at 20 °C in line with normal experience. For the W/(C+S) = 0.30 mixtures however, the heat curing does not lead to reduced compressive strength. This finding is discussed in [5] and [6].

In Table 7 and Figures 8 -11 results from the calorimetric ice formation measurements are given.

Table 7. Ice formation by calorimetry. Dried/Resaturated specimens.

		03 - 00		03 - 08	
		20 °C	60 °C	20 °C	60 °C
1)We (CAL)	g/g_{SSD}	0.049	0.048	0.051	0.058
COOL	g/g_{SSD}	0.00184	0.0033	0.00123	0.00120
Ice to - 8 °C % of	% of W_e	3.8 %	6.9 %	2.4 %	2.1 %
HEAT	g/g_{SSD}	0.0103	0.0139	0.0117	0.0169
Total Ice	% of W_e	22.0 %	29.0 %	22.9 %	29.1 %

1) Total evaporable water at test

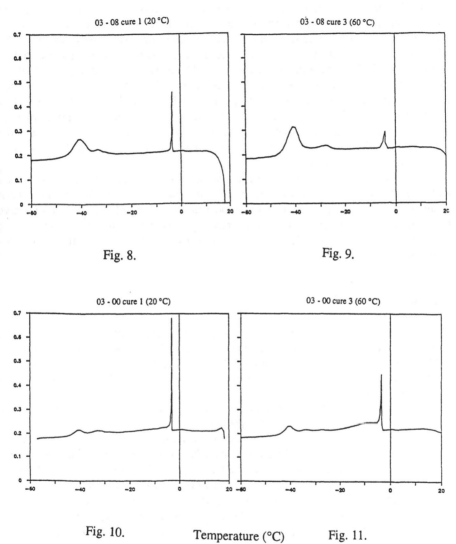

Fig. 8.   Fig. 9.

Fig. 10.   Temperature (°C)   Fig. 11.

Calorimetric ice formation: Appearent heat capacity vs. temperature

The figures show different ice formation patterns with and without SF: 03-00 shows a marked freezing peak around - 4°C, and ice formation continues on further cooling - at a higher rate for the 60 °C cure. The amounts of ice formed to - 8 °C correspond to 3.8 and 6.9 % of the evaporable water for 20 and 60 °C curing, respectively (Table 7). For 03-08 peaks also exist at - 4 °C, but there is insignificant ice formation as cooling proceeds to - 20 °C. The amounts of ice to - 8 °C correspond to 2.4 and 2.1 % of the evaporable water for 20 and 60 °C cure, respectively. The total amounts of ice formed to - 55 °C are calculated from the heating curves and given in Table 7. There is a clear temperature effect for both mixtures; 60 °C curing leads to increased ice formation, implying a more coarse pore structure. The major effect of SF is clearly that most of the ice formation takes place below - 20 °C, as seen earlier [9], and most likely plays a less important role for frost resistance.

The present results show that there is a qualitative correlation between ice formation and salt/frost scaling: elevated temperature curing increases both. The same has been found as a consequence of dry/rewet treatment before testing [1, 4]. SF has a positive effect on the salt/frost scaling after both types of pretreatment. The results also indicate that high strength concrete does not necessarily need air entrainment to be salt/frost resistant.

5 Conclusions

Heat curing of normal strength concrete results in a reduced salt/frost scaling resistance compared to concrete cured at 20 °C. For non-airentrained concretes with W/(C+S) = 0.45 and S/(C+S) = 0 and 0.08, curing at 60 °C leads to 4 to 8 times higher scaling than for concrete cured at 20 °C. For the same concrete mixtures cured at 40 °C the increase in salt/frost scaling is lower.

For a non-airentrained high strength concrete with W/(C+S) = 0.30 without SF, 60 °C heat cure also may lead to increased scaling.

The two non-airentrained high strength concretes with W/(C+S) = 0.30 and 8 % SF showed excellent salt/frost scaling resistance after heat curing (60 °C), even after a dry/rewet - pre treatment and 280 freeze/thaw cycles. The use of 8 % SF leads to reduced scaling for both W/(C+S) - ratios and all heat curing procedures up to 60 °C.

Ice formation down to - 20 °C were negligible for concretes with w/(c+s) = 0.30 and s/(c+s) = 0.08 both for 20°C and 60 °C curing. Concrete with w/c = 0.30 without SF showed significant ice-formation after 60 °C curing, but not after 20 °C curing. The results as a whole confirm the finding [4], that there is at least a qualitative correlation between ice formation and frost/salt scaling of concrete.

6 References

1. Sellevold E.J., Bager D.H.(1985) Dansk Betonforening Publ. Nr.22, pp.47-79
2. Bager D.H. and Sellevold E.J.(1986) Cem. and Conc. Res., Vol 16, pp. 835 - 844
3. Villadsen J.(1989) Diploma thesis, DTH-BML, Denmark (In Danish).
4. Sellevold E.J., Farstad T.(1991) Nordic Concrete Research publ. no 10, pp. 121 - 138, Norsk Betongforening, Oslo.
5. Laamanen P.H., Johansen K., Kyltveit B.P. and Sellevold E.J.(1992) ACI SP 132 Vol. 2 pp.1045-1059
6. Lindgård J. and Sellevold E.J. (1993) Proceedings Utilization of high strength concrete Vol. 2 Int. Symp., Lillehammer, Norway, pp. 810 - 821.
7. Petersson P.E.(1988) Proceedings Int. Sem. Chalmers Techn. Univ., Gøteborg 1986. Publication DI:1988, Swedish Council for Building Research.
8. Sellevold E.J. (1986) Report O 1731, Norwegian Building Research Institute, Oslo, (In Norwegian)
9. Sellevold E.J., Bager D.H., Klitgaard Jensen E., and Knutsen T.(1982) The Norwegian Institute of Technology, Report BML 82.610 pp. 19-50

Acknowledgement

The work reported here is part of the Norwegian national project HIGH STRENGTH CONCRETE: MATERIALS DESIGN. The project is supported by The Royal Norwegian Council for Scientific and Industrial Research and a major part of the concrete industry, with NORCEM A/S as project leader.

7

Freeze-deicing salt resistance of concretes containing blast-furnace slag-cement

J. STARK and H.-M. LUDWIG
F.A. Finger-Institut für Baustoffkunde, Weinmar, Germany

Abstract
Whereas blast-furnace cement concretes have proved successful in structures subjected to freeze-thaw attack, their use in structures subjected to freeze-deicing salt attack is still a problem. The present paper discusses the causes of this controversy. It was found out that the freeze-deicing salt resistance of blast-furnace cement concretes is closely related to the carbonation of the surface area. The carbonation of blast-furnace cement concretes does not only lead to an increase in capillary porosity but also to metastable calcium carbonates soluble in NaCl. Based on the results of investigations ways of improving the freeze-deicing salt resistance of blast-furnace cement concretes are proposed.
Keywords: aragonite; calcite; carbonation; degree of hydration; freeze-deicing salt resistance blast furnace cement; frost resistance; porosity; vaterite.

1 Introduction

Whereas cements rich in granulated blast-furnace slag have proved successful in structures subjected to frost attack such as dams, their use in structures which are subjected to a very heavy attack by deicing salts (e.g. in concrete road decks) is restricted. According to DIN 1045 only grade 45 blast-furnace cements may be used under heavy freeze-deicing salt attack, i.e. blast-furnace cements with a medium content of granulated blast-furnace slag. That means that the use of grade 35 blast-furnace cements, which are richer in granulated slag (as a rule > 60 %), is not permitted in these cases. Practical experience shows, however, that concretes rich in

Freeze-Thaw Durability of Concrete. Edited by J. Marchand, M. Pigeon and M. Setzer.
Published in 1997 by E & FN Spon, 2–6 Boundary Row, London SE1 8HN, UK.
ISBN 0 419 20000 2.

granulated slag without air-entrainment were successfully employed in sewage treatment plants.

The standard requirements of cement used in concretes with a high freeze-thaw and/or freeze-deicing salt resistance reflect the present state of art in this field.

There seems to be general agreement that concrete containing blast-furnace cement with a granulated slag content of more than 60 % may have a high frost resistance (proper curing provided), sometimes even higher than comparable portland cement concretes [1-3].

There is more controversy over the freeze-deicing salt resistance of blast-furnace cements. In reference to laboratory tests many authors hold the view that freeze-deicing salt resistance will be reduced by the use of blast-furnace cements [4-6]. Bonzel/Siebel [7] found that the freeze-deicing salt resistance of concretes rich in granulated blast-furnace slag (content > 60 %) cannot be improved by air-entraining agents in the same way as comparable portland cement concretes can. Similar results were reported by Hilsdorf [8]. He found that the freeze-deicing salt resistance of blast-furnace cement concretes is not definitely improved by the addition of air-entraining agents, even if this leads to highly favourable air voids spacing factors of 0.12 mm and 0.04 mm respectively in the hardened concrete. Only few authors achieved results which indicate that the freeze-deicing salt resistance of blast- furnace cement concretes may be improved by air-entrainment in the same way as portland cement concretes [9,10].

So far, no definite reasons have been found for the high frost resistance of concretes rich in granulated blast-furnace slag on the one hand and their low freeze-deicing salt resistance in laboratory tests on the other. Furthermore, it remains unclear why in some cases an adequate air void system in air-entrained blast-furnace cement concretes is of no effect, whereas the freeze-deicing salt resistance of portland cement concretes with comparable air void parameters is considerably improved.

This paper presents the results of several years research work in the field of freeze-thaw and freeze-deicing salt resistance and will help clarify the contradictions mentioned above. The observations on blast-furnace slag cement refer exclusively to grade 35 cement - at least 35 N/mm^2 after 28 days - i.e. blast-furnace cement 35 L according to DIN 1164 (as a rule, the granulated slag content of grade 35 blast furnace cements is >55%).

The investigations make a contribution to the problem of freeze-thaw and freeze-deicing salt resistance of concretes made of blast-furnace cements rich in granulated slag. New findings were made in particular on the effect of the degree of hydration and the carbonation. The test results were the starting point for the search for ways of improving the freeze-deicing salt resistance of blast-furnace cement concretes.

2 Frost resistance of blast-furnace cement concrete

The problem of frost attack is a problem of the volume, i.e. in addition to surface scaling the interior microstructure may also be destroyed. This fact is reflected in the frost resistance testing criteria of concretes, which consider not only the change in mass and/or the amount of scaling but also characteristics describing the state of the microstructure within the concrete (decrease of the dynamic modulus of elasticity, of the ultrasonic speed, of the compressive strength etc.). Most of the freeze-tests in the investigations described below were made according to procedure A of TGL 33433/06. During the freeze-thaw cycles (+20°C/-20°C) the concrete cubes (edge length 10 cm) are completely immersed in water, both in the freezing process and in the phase of thawing. A regular test consists of 100 freeze-thaw cycles. The damage is measured by means of the decrease in ultrasonic speed.

The degree of hydration (which differed depending on: the quality of the blast-furnace slag, the fineness of the cement, and the differences in pore size distribution) had a strong influence on the frost resistance of mortars and concretes in laboratory tests. For determining the relationship between the degree of hydration and frost resistance we used the above mentioned cements for the production of concretes without air-entrainment, which were afterwards tested for their frost resistance (cement content = 350 kg/m^3), water-cement ratio = 0.5). By means of a plasticiser a similar consistency of all concretes could be achieved, which was in the range of standard consistency (slump = 42 to 48 cm) according to German Standards (DIN 1045). Also the fresh air content was in the very narrow range of 0.9 to 1.3 % by volume. Consequently, any influence of different air contents or slump on the frost test results was excluded. As may be seen in Fig. 1, there exists a linear relationship between the frost resistance and the degree of hydration of blast-furnace cement.

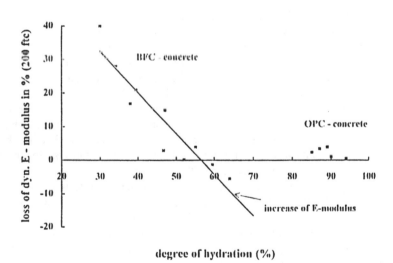

Fig. 1. Dependence of frost resistance (TGL 33433/06-A) on the degree of hydration of the cements; concretes: water-cement ratio = 0.5,
cement content = 350kg/m^3, without air-entraining agents (a.e.a.)

Compared to portland cement blast-furnace cements may be subdivided into three ranges, which depend on pore size distribution:

$\alpha_H < 50\,\%$ -   the lower degree of hydration leads to a highly capillary-porous microstructure, the frost resistance is lower than that of portland cement mortars and concretes

$\alpha_H = 50\,\%$ -   the low degree of hydration may be compensated for by the denser microstructure, the frost resistance corresponds to that of portland cement mortars and concretes

$\alpha_H > 50\,\%$ -   the formation of a very dense microstructure results in a frost resistance which is higher than that of portland cement mortars and concretes

The studies of frost resistance show that a general assessment of blast-furnace cement concrete is not possible. The frost resistance differs greatly, depending on the cement quality, particularly on the quality of the granulated blast-furnace slag, and the curing.   In this context it has to be emphasised that all blast-furnace cements examined contained slag which was in accordance with the requirements of DIN 1164 (1990) for an F2 value of ≥ 1. However, an F2 value of at least 1.5 was required to achieve a high frost resistance.

3 Freeze-deicing salt resistance of blast-furnace cement concrete

As already stated, the freeze-deicing salt resistance of blast-furnace cement concretes is generally considered to be worse than that of comparable portland cement concretes. The fact that in some cases the air-entraining agents, which are specified in blast-furnace cement concretes under heavy freeze-deicing salt attack, did not sufficiently increase the freeze-deicing salt resistance led to the exclusion of cement rich in granulated blast-furnace slag (grade 35) from these applications.

Again there was reason to suppose that the degree of hydration is responsible for the different freeze-deicing salt resistance of blast-furnace and portland cement concretes. To verify this assumption the concretes without air-entrainment described in the passage above were not only tested for their freeze-thaw resistance, but also for their freeze-deicing salt resistance by means of the CDF-procedure.

In contrast to the pure frost test the results of the freeze-deicing salt tests showed no relationship between the freeze-deicing salt resistance and the degree of hydration (Fig. 2). This becomes apparent if it is taken into account that, unlike pure frost attack, freeze-deicing salt attack causes damage primarily to the surface of the concrete. Studies showed that, even with great surface scaling, no significant interior damage of the concrete could be observed [11]. Therefore almost all procedures of determining the freeze-deicing salt resistance of concrete are confined to determining the loss in weight and/or the amount of scaled material. Freeze-deicing salt tests conducted using the CDF procedure at the HAB Weimar (University of Architecture and Building) resulted in amounts of scaling between 90 g/m^2 and 6000 g/m^2 after 28 freeze-thaw cycles, depending on the quality of the concretes. If, assuming a bulk density of 2300 kg/m^3 various scalings are expressed in terms of an average depth of scaling the result will be a layer of 0.04 mm to 2.6 mm in depth. The above

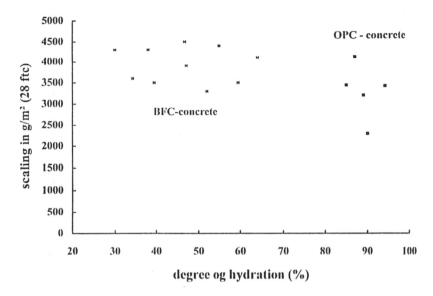

Fig. 2. Dependence of freeze-deicing salt resistance (CDF test) on the degree of hydration of the cements concretes: water-cement ratio = 0.5, cement content = 350 kg/m^3, without air-entraining agents (a.e.a.).

considerations lead to the conclusions that the degree of hydration and the state of the microstructure of the concrete as a whole are of minor importance to whether a concrete exhibits sufficient freeze-deicing salt resistance or not. The properties of a thin concrete surface layer are decisive.

To understand the different freeze-deicing salt resistance of blast-furnace and portland cement concretes, it is important to know in which stage of freezing differences will occur. Whereas in tests for frost resistance under water the scaling of both blast-furnace and portland cement concrete is linear (Fig. 3), heavy initial scaling is observed in tests for the freeze-deicing salt resistance of concrete containing cements rich in granulated blast-furnace slag (content ≥ 60 %) in 3 % NaCl solution (Fig. 4). This heavy initial scaling occurs both in concretes with and without air-entrainment and is responsible for the fact that in many cases blast-furnace cement concretes show a lower freeze-deicing salt resistance than comparable portland cement concretes and that air-entraining agents are less effective in cements rich in granulated blast-furnace slag. After the initial scaling in the first 4 - 8 freeze-thaw cycles the curve changes appreciably and - as a rule - indicates in the second phase a similar or even lower intensity of damage compared to portland cement concretes. The reasons for the heavy initial scaling of blast-furnace cement concretes are not yet known. As the material scaled at the beginning of the damage is identical to the surface layer of the concretes it may be assumed that the initial scaling is influenced by the carbonation of the concrete.

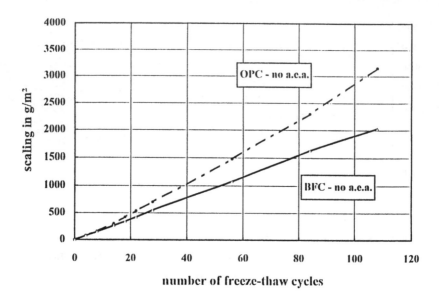

Fig. 3. Typical courses of scaling of portland cement concrete (OPC) and blast-furnace cement concrete (BFC) with grade 35 cement without air-entraining agents (a.e.a.) under frost attack (CF test)

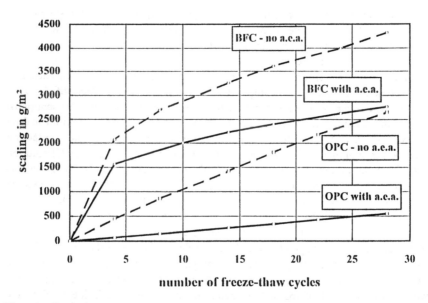

Fig. 4. Typical courses of scaling of portland cement concrete (OPC) and blast-furnace cement concrete (BFC) with grade 35 cement with and without air-entraining agents (a.e.a.) under freeze-thaw salt attack (CDF test)

3.1 Tests for the influence of carbonation

In order to find out whether the heavy initial scaling of blast-furnace cement concretes is related to the carbonation, the depth of carbonation of different blast-furnace cement concretes prior to freezing was compared to the calculated depth of scaling at the break point of the curve of scaling. The results of these studies (Fig. 5) show conclusively that the decrease in the intensity of damage (break point of scaling curve) occurs precisely when the damage proceeds from the carbonated to the non-carbonated zone of the blast-furnace cement concretes. Furthermore, the scaled material was subjected to thermoanalysis and the amount of calcium carbonate in the material was determined by means of the DTA/TG curve for each phase of freezing. The comparison of the calculated amounts of calcium carbonate with the behaviour of the curves of scaling shows (Fig. 6), in accordance with the studies into the depths of carbonation, that exactly at the break point of the curve of scaling the content of carbonate in the scaled material decreases considerably. That means that heavy scaling will occur only in the carbonated surface area, whereas the non-carbonated core shows sufficient resistance to freeze-deicing salt attack.

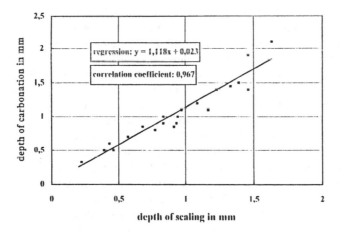

Fig. 5. Relationship between depth of initial scaling (calculated from amount of scaling at the point of discontinuity of the curve of scaling) and depth of carbonation of blast-furnace cement concretes with and without a.e.a.

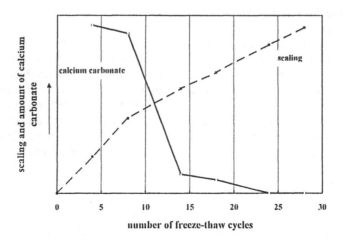

Fig. 6. Comparison between the curve of scaling of a blast-furnace cement concrete under freeze-deicing salt attack and the amount of calcium carbonate in the scaled material

Since carbonation has a strong influence on the initial scaling of blast-furnace cement concretes, it may be supposed that with concretes stored in lower CO_2 concentrations less scaling would occur during freezing and that the damage would be greater in higher CO_2 concentrations. In order to verify this assumption a blast-furnace cement concrete was produced with a granulated slag content of 65 %, no air-entrainment, a water-cement ratio of 0.5, a cement content of 350 kg/m^3,and stored in water for one week. After the water storage period the specimens were separated and stored for three weeks under different ambient conditions at a temperature of 20 °C and a relative humidity of 65 %. According to the CDF procedure specifications some specimens were air-stored. The other specimens were stored without CO_2 under nitrogen or in an increased CO_2 concentration (3 % CO_2) in a carbonising chamber. At the beginning of the CDF test the depth of carbonation was determined. The usual indicator technique could not be applied to the specimens stored in nitrogen because of the small depth of carbonation, and an automatic image analysis with thin ground sections had to be used instead. The following depths of carbonation were determined:

storage in nitrogen - 0.3 mm
storage in air   - 1.5 mm
storage in CO_2 - 12.5 mm

The results of the subsequent CDF test differed greatly depending on the kind of pre-storage (Fig. 7).

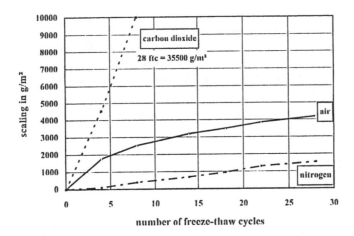

Fig. 7. Freeze-deicing salt resistance of blast-furnace cement concretes without a.e.a. - stored under different conditions (BFC grade 35; water-cement ratio = 0.5; cement = 350 kg/m³)

Air storage

After 28 freeze-thaw cycles the average scaling of the air-stored concrete specimens amounted to about 4000 g/m². The course of scaling confirmed the statement that the progressive initial scaling during the first 4 - 8 freeze-thaw cycles is followed by a linear course of damage of a lower intensity.

CO_2 storage

The scaling of the concretes stored in increased CO_2 concentration amounted to about 35000 g/m² after 28 freeze-thaw cycles, which was the worst freeze-deicing salt resistance of all concretes. The typical break point in the scaling process curve is shifted to later freeze-thaw cycles. It occurs after the 24th freeze-thaw cycle with an amount of scaling of about 30000 g/m²,which for a bulk density of 2300 kg/m³ would correspond to a depth of scaling of about 13 mm. (depth of carbonation = 12.5 mm)

Nitrogen storage

With an amount of scaling of about 1500 g/m² the concretes stored in nitrogen exhibited the best freeze-deicing salt resistance. In contrast to the air-stored specimens these concretes showed no progressive initial scaling in the CDF test, and the course of scaling was linear from the beginning. The increase in scaling corresponded closely to that of the air-stored specimens in their second linear phase of damage.

Similar tests were subsequently made on blast-furnace cement concretes with air-entraining agents. Under standard storage - 1 week in water, 3 weeks in the air at 20 °C and 65 % relative humidity - initial scaling was hardly reduced by the air-entrainment compared to 0-concrete. In the second phase of damage, however, the amount of scaling was reduced (Fig. 8). The specimens stored in nitrogen showed no initial scaling at all.

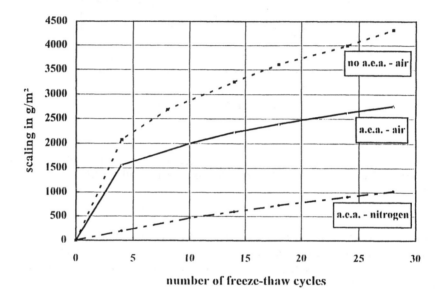

Fig. 8. Freeze-deicing salt resistance of blast-furnace cement concretes with and without a.e.a. - stored under different conditions (BFC grade 35; water-cement ratio = 0.5; cement = 350 kg/m3)

In order to study the causes of the adverse effect of carbonation on the initial scaling of blast-furnace cement concretes, basic investigations into cement pastes (phase-analytical investigations) and concretes (investigations into the microstructure) were carried out.

Differences between blast-furnace and portland cement concrete became apparent through the effect of carbonation on the microstructure. While carbonation led to a slight densification of the microstructure in portland cement concrete, the microstructure of blast-furnace cement concrete became coarser. In fact, the total porosity decreased also in this case, at the same time the portion of capillary pores increased. Nevertheless, it occured after an even smaller carbonation than that of portland cement concrete. The coarser microstructure of blast-furnace cement concretes due to carbonation has already been stated by other authors [12,13]. In general it is explained by the fact that (in contrast to portland cement concrete) blast-furnace cement concrete carbonates mainly in the C-S-H phases, where a highly porous silica gel is formed. The increase of capillary pores due to carbonation, however, cannot be the main reason for blast-furnace cement concretes behavior under freeze-deicing salt attack, since in this case also under frost attack a higher intensity of damage in the carbonated surface area should have been expected. In contrast to the freeze-deicing salt attack, no break point is observed for the pure freeze-thaw attack under water when the damage proceeds from the carbonated to the non-carbonated zone of the concrete (Fig. 3).

The different course of scaling of blast-furnace cement concretes under frost attack in water and in NaCl solution indicates that there is a chemical cause.

Phase-analyses of the carbonated surface area (by the XRD) prior to freezing showed in blast-furnace cement that in addition to calcite two other $CaCO_3$ modifications, aragonite and vaterite occur in considerable amounts (Fig. 9).

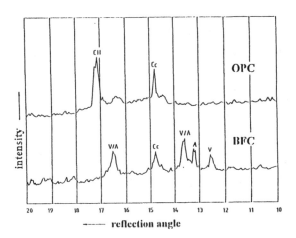

Fig. 9. XRD of the carbonated surface area of portland cement pastes and blast-furnace cement pastes; CH = portlandite; Cc = calcite; A = aragonite; V = vaterite

In portland cement, however, only calcite could be detected. The causes of the occurrence of different calcium carbonate modifications in blast-furnace cement are to date not completely clear. Possibly the different carbonate phases found in portland cement and blast-furnace cement are due to the fact that $Ca(OH)_2$ carbonates to calcite whereas the C-S-H phases or AFm and AFt phases carbonate to aragonite and vaterite [14,15]. Thermoanalyses showed that in all cases the decarbonation temperature of the $CaCO_3$ was between 800 and 900 °C. During the DTA test heating period the metastable $CaCO_3$ modifications are transformed to calcite at about 300 - 400 °C (exothermic peak - Fig. 10). Examinations of the scaled material and the damaged surface area after the freeze-deicing salt attack showed that large amounts of $CaCO_3$ had already decomposed at about 650 °C (Fig. 10). It was proved by mass spectrometry that this reaction is indeed a decompositon of $CaCO_3$. According to Cole/Kroone [16] a lower decomposition temperature indicates badly crystallised

calcite. X-ray analysis showed that the new formation of this calcite is accompanied by an intense reduction of aragonite and vaterite. These observations coincide with the existing publications on stability and transformation of calcium carbonate modifications. According to them the solubility constants of the metastable modifications are larger than those of the calcite, and these differences increase considerably in chloride solutions [17,18]. From the order of solubilities
K_L (calcite) < K_L (aragonite) < K_L (vaterite) it follows that a solution which is saturated with respect to vaterite and aragonite is always over-saturated with respect to calcite. Under certain conditions the decomposition of vaterite and aragonite is followed by formation and growth of calcite nuclei.

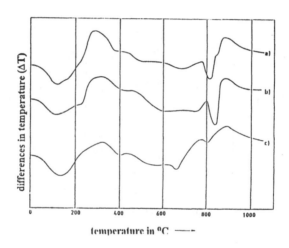

Fig. 10. DTA of blast-furnace cement pastes
a) prior to freezing and thawing
b) after freezing and thawing in water
c) after freezing and thawing in 3% NaCl solution

Based on the test results the following hypothesis about the heavy initial scaling of concretes rich in granulated blast-furnace slag may be proposed: The metastable modifications of $CaCO_3$ - vaterite and aragonite - are dissolved by the combined attack of frost and chloride. This leads to an extensive erosion of the concrete surface. In the deeper carbonated layers the dissolution of vaterite/aragonite may lead to the new formation of calcite in the microstructure. The change in volume implied may further aggravate the damage initiated by the dissolution processes. After the entire carbonated layer has been scaled in this way the speed of damage decreases appreciably - a clear break point is to be observed in the curve of scaling (Fig. 4).

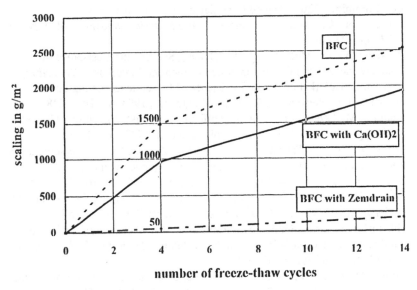

number of freeze-thaw cycles

Fig. 11. Initial scaling of blast-furnace cement concretes without a.e.a. - (BFC grade 35; water-cement ratio = 0.5; cement = 350 kg/m^3)
a) BFC
b) BFC with 10 % Ca(OH)$_2$
c) BFC with Zemdrain

6 Conclusions

The investigations show that the freeze-deicing salt resistance of concretes rich in granulated blast-furnace slag is strongly influenced by the carbonation of the surface layer. The coarser microstructure due to carbonation as well as the existence of metastable carbonates (vaterite, aragonite) lead to a freeze-deicing salt resistance of the carbonated layer which is lower than that of the concrete core. In order to achieve a high freeze-deicing salt resistance it is therefore necessary to minimise the carbonation of these concretes. In addition to well-known methods such as a low water-cement ratio and proper curing, draining formwork materials is particularly suitable.

7 References

1. Siebel, E., and Sprung, S., "Einfluß des Kalksteins im Portlandkalkstein-zement auf die Dauerhaftigkeit von Beton." *Beton*, Vol. 41, No. 3, 1991, pp. 113 - 117.
2. Stark, J.; Ludwig , H.-M., and Müller, A., Der Zusammenhang von Hydratationsgrad und mörteltechnischen Kenngrößen. *Silikattechnik*, Vol. 42, No. 1, 1991, pp. 28 - 30.

3.   ACI Committee 226, "Ground Granulated Blast - Furnace Slag as a Cementitious Constituent in Concrete." *ACI Materials Journal*, Vol. 4, No. 4, 1987, pp. 327 - 342.

4.   Bilodeau, A., and Malhotra, V.M., "Deicing Salt Scaling Resistance of Concrete Incorporating Supplementary Cementing Materials." *Proceedings*, International Workshop on the Resistance of Concrete to Scaling due to Freezing in the Presence of Deicing Salt, Quebec, 1993, pp. 189 - 228.

5.   Schmidt, M., "Zement mit Zumahlstoffen - Leistungsfähigkeit und Umwelt-belastung." *Zement - Kalk - Gips*, Vol. 45, No. 2, 1992, pp. 64- 69.

6.   Schorr, K., "Frost - Tausalz - Widerstand von Zementstein aus verschiedenen Zementarten und mit unterschiedlichen Beimengungen von Flugasche." *Betonwerk + Fertigteil - Technik*, Vol. 50, No. 1, 1983, pp. 16 - 21.

7.   Bonzel, J., and Siebel, E., "Neuere Untersuchungen über den Frost - Tausalz - Widerstand von Beton." *Beton*, Vol. 27, No. 4, 1977, pp. 153 - 157.

8.   Hilsdorf, H.K., and Günther, M., "Einfluß der Nachbehandlung und Zementart auf den Frost - Tausalz - Widerstand von Beton." *Beton- und Stahlbetonbau*, Vol. 81, No. 3, 1986, pp. 57 - 62.

9.   Luther, M.D., "Scaling Resistance of Ground Granulated Blast Furnace Slag Concretes." *Proceedings*, 3rd International CANMET/ACI - Conference "Durability of Concrete", Nice, 1994, pp. 47 - 64.

10.   Virtanen, J., "Mineral By-Products and Freeze - Thaw Restistance of Concrete." *Nordic Concrete Research*, Vol. 3, 1984, pp. 191 - 209.

11.   Siebel, E., "Frost- und Frost - Tausalz - Widerstand. Beurteilung mittels Würfel-verfahren." *Beton*, Vol. 42, No. 9, 1992, pp. 496 - 501.

12.   Manns, W., and Zeus, K., "Zur Bedeutung von Zuschlag und Zement für den Frost - Taumittel - Widerstand von Beton." *Straße und Autobahn*, Vol. 30, No. 4, 1979, pp. 167 - 173.

13.   Bier, T.A., "Karbonatisierung und Realkalisierung von Zementstein und Beton." *Diss.*, Universität Karlsruhe, 1987.

14.   Suzuki, K., and Nishikawa, T., "Formation and Carbonation of C-S-H in water." *Cement and Concrete Research*, Vol. 15, 1985, pp. 213 - 224.

15.   Nishikawa, T.; Suzuki, K.; Ito, S.; Sato, K., and Takebe, T., "Decomposition of synthesized ettringite by carbonation." *Cement and Concrete Research*, Vol. 22, 1992, pp. 6 - 14.

16.   Cole, W.F., and Kroone, B., "Carbon Dioxid in Hydrated Portland Cement." *Journal ACI*, Vol. 31, No. 6, 1960, pp.1275 - 1295.

17.   Schröder, F., "Vaterit, das metastabile Calciumcarbonat, als sekundäres Zementstein-mineral." *Tonindustriezeitung*, Vol. 86, No. 10, 1962, pp. 254 - 260.

18.   Richter, A., " Ein Beitrag zur Fällung und Phasenumwandlung von Calciumcarbonat." *Diss.*, Bergakademie Freiberg, 1993.

8

Deicing salt scaling resistance of concrete incorporating supplementary cementing materials: CANMET research

A. BILODEAU and V.M. MALHOTRA
CANMET, Natural Resources Canada, Ottawa, Canada

Abstract
This paper presents the results of four CANMET investigations dealing with the deicing salt scaling resistance of concrete incorporating supplementary cementing materials. The supplementary cementing materials investigated included silica fume, blast-furnace slag and fly ash. The scaling resistance of the superplasticized concrete incorporating high volumes of fly ash was also determined.

In general, the incorporation of supplementary cementing materials affected adversely its deicing salt scaling resistance. This effect ranged from being very marginal to considerable depending upon the type and percentage of the supplementary cementing materials used. Concretes incorporating 8% silica fume or up to 30% fly ash performed satisfactorily in the ASTM deicing salt scaling test C 672. In particular, the performance of fly ash concrete was noticeably improved by the use of membrane curing. The slag concretes showed considerably more scaling than the reference concretes, and in general, their surfaces were rated as moderately scaled. All high-volume fly ash concretes performed relatively poorly in the scaling test. Their poor performance is possibly related to the unsatisfactory quality of their air-void parameters, principally at the surface of the specimens.
Keywords: Blast-furnace slag, Deicing salt, Fly ash, High-volume fly ash, Membrane curing, Moist curing, Scaling, Silica fume

1 Introduction

Since the 1970's, CANMET has been engaged in research on concrete containing supplementary cementing materials such as fly ash, blast-furnace slag and silica fume [1-5]. When used in proper proportions and when correct concrete making practices

Freeze-Thaw Durability of Concrete. Edited by J. Marchand, M. Pigeon and M. Setzer.
Published in 1997 by E & FN Spon, 2–6 Boundary Row, London SE1 8HN, UK.
ISBN 0 419 20000 2.

are followed, these materials can be used to produce concrete having excellent mechanical properties and durability characteristics.

Recently, CANMET has developed high-volume fly ash concrete, a structural concrete incorporating high volumes of low-calcium (ASTM Class F) fly ash [6-9]. Typically, in this type of concrete, the water and cement contents are kept low at about 115 and 155 kg/m³ of concrete, respectively, and the proportion of low-calcium fly ash in the total cementitious materials content ranges from 55 to 60%. A high degree of workability is obtained by the use of a superplasticizer. Numerous investigations performed at CANMET and other institutions have shown this concrete to have excellent structural properties.

One of the concerns about the use of supplementary cementing materials in concrete is the resistance of the concretes incorporating these materials to the combined action of freezing and thawing cycling and deicing salts. Some published data indicate that these concretes scale more than concretes made with portland cement alone [10-12]. The concern is even greater for high-volume fly ash concrete, considering the very low amounts of cement and the large proportion of fly ash used in this type of concrete.

This paper is a review of four CANMET investigations dealing with the deicing salt scaling resistance of concrete incorporating silica fume or blast-furnace slag or fly ash; the scaling resistance of the high-volume fly ash concrete was also determined [13-16]. Some tests to determine the factors affecting the scaling resistance of concrete were also performed, and are reported here.

2 Deicing salt scaling resistance of silica fume concrete

This study is divided into two parts. In the first part (Part A), eighteen 0.06 m³ concrete mixtures were made. Half of them incorporated 8 per cent silica fume as a replacement by mass for cement. The water-to-(cement + silica fume) ratio ranged from 0.40 to 0.65. All mixtures were air-entrained, some contained a superplasticizer, and test slabs were cast for determining the scaling resistance of the concrete surfaces when exposed to deicing salts. Also, from each mixture a number of cylinders and prisms were cast for testing in compression and flexure, and for determining the resistance of concrete to repeated cycles of freezing and thawing. Sawn sections of the test prisms were used for determining the air-void parameters of the hardened concrete.

In the second part (Part B), two mixtures from part A were repeated; these were non-superplasticized mixtures made with and without silica fume, using a water-to-cementitious materials ratio of 0.60. From each mixture, a number of slabs were cast for the scaling resistance of the concrete. In this case, the test included different methods of preparation, curing and testing of the specimens. In addition, cylinders and prisms were cast for testing in compression and flexure. Once again, sawn sections of the prisms were used for determining the air-void parameters of the hardened concrete.

2.1 Materials
The concrete mixtures were prepared using the following materials.

2.1.1 Cement
Normal portland cement, ASTM Type I was used. Its physical properties and chemical analysis are given in Table 1.

2.1.2 Silica fume
Silica fume from a source in Quebec was used. Its physical properties and chemical analysis are also given in Table 1.

Table 1. Physical properties and chemical analysis of cement and silica fume

	Portland cement	Silica fume
Physical properties		
Time of set, (min.)		
initial:	130	
final:	235	
Fineness;		
45 μm (passing), (%)	82.7	
surface area, Blaine, (m^2/kg)	366	
surface area, nitrogen adsorption, (m^2/kg)		20,000
Soundness, autoclave expansion, (%)	0.13	
Compressive strength of 51-mm cubes, (MPa)		
3 d	22.6	
7 d	27.0	
28 d	32.9	
Chemical analysis, (%)		
SiO_2	20.14	94.00
Al_2O_3	5.40	0.06
Fe_2O_3	2.38	0.03
CaO	62.04	0.50
MgO	2.62	1.10
Na_2O	0.24	0.02
K_2O	1.12	0.20
SO_3	3.87	0.50
Loss on ignition	----	2.50
Bogue potential compound composition, (%)		
C_3S	48.7	
C_2S	21.0	
C_3A	10.3	
C_4AF	7.2	

2.1.3 Aggregates
The coarse and fine aggregates consisted of minus 19 mm crushed dolomitic limestone and local natural sand respectively. To keep the grading uniform for each

mixture, both the fine and coarse aggregates were separated into different size fractions that were then recombined to a specific grading. The grading and the physical properties of both aggregates are given in Tables 2 and 3.

Table 2. Grading of aggregates

Coarse aggregates		Fine aggregates	
Sieve size, (mm)	Cumulative percentage retained	Sieve size, (mm)	Cumulative percentage retained
19	33.3	4.75	0.0
9.5	66.6	2.36	10.0
4.75	100.0	1.18	32.5
		1.40	57.5
		0.300	80.0
		0.150	94.0
		pan	100.0

Table 3. Physical properties of aggregates

	Coarse aggregates	Fine aggregates
Specific gravity	2.68	2.70
Absorption, (%)	0.80	1.1

2.1.4 Air-entraining admixture
A sulphonated hydrocarbon type air-entraining admixture was used in all the mixtures.

2.1.5 Superplasticizer
A sulphonated naphthalene formaldehyde condensate type superplasticizer was used. It is usually available as a 42 per cent solids aqueous solution with a density of 1200 kg/m^3, and is dark brown in color. The chloride content is negligible.

2.2 Mixture proportions
In Part A, eighteen mixtures divided into two series were made. The first series included the use of a superplasticizer, and had the water-to-(cement+silica fume) ratio (W/(C+SF)) ranging from 0.40 to 0.65, whereas, the second series was made without the superplasticizer and with W/(C+SF) ranging from 0.55 to 0.65. In Part B (mixtures 19 and 20), the mixtures of the second series of Part A having a W/(C+SF) of 0.60 were repeated.

For each W/(C+SF), two concrete mixtures were made, one reference mixture (without silica fume) and a mixture incorporating 8 per cent of silica fume as a replacement by mass for cement. As far as possible, the slump of all mixtures was maintained at 100 ± 25 mm, and the air content was kept at 6.5 ± 0.5%. Mixing was carried out in a laboratory counter-current pan mixer with the simultaneous addition of silica fume and cement. The mixture proportion and the properties of the freshly-mixed concrete are given in Table 4.

Table 4. Mixture proportions and properties of fresh concrete

Part	Mix No.	W/(C+SF)	Cement, (kg/m³)	Silica Fume, (kg/m³)	SP,* (L/m³)	AEA,** (mL/m³)	Slump, (mm)	Air Content, (%)
	1	0.40	343	0	2.6	380	115	6.6
	2	0.40	316	27	3.9	1135	75	7.0
	3	0.45	307	0	1.8	230	100	6.3
	4	0.45	279	24	3.3	525	150	7.4
	5	0.50	276	0	2.2	305	125	5.9
	6	0.50	250	22	2.3	350	115	7.0
	7	0.55	245	0	1.8	305	100	6.7
	8	0.55	225	19	2.0	300	85	6.2
A	9	0.60	223	0	1.8	230	100	6.2
	10	0.60	203	18	1.5	240	75	7.0
	11	0.65	203	0	1.8	290	110	6.1
	12	0.65	186	17	1.8	225	100	6.8
	13	0.55	270	0	0	170	95	6.5
	14	0.55	256	23	0	200	70	6.8
	15	0.60	241	0	0	135	115	6.9
	16	0.60	231	20	0	165	70	6.8
	17	0.65	218	0	0	120	110	6.5
	18	0.65	208	18	0	135	90	6.7
B	19	0.60	240	0	0	135	110	6.5
	20	0.60	228	20	0	145	90	6.7

* Superplasticizer
** Air-entraining admixture

2.3 Preparation and casting of test specimens
Four 152x305 mm cylinders, six 76x102x387 mm prisms and two 280x300x75 mm slabs were cast from each mixture in Part A of the study. In Part B, four 102x203 mm cylinders, two 76x102x406 mm prisms and eight 280x300x75 mm slabs were cast from each mixture. All specimens were cast in two layers, each layer being compacted using an internal vibrator for the 152x305 mm cylinders, and a vibrating table for the other specimens.

The finishing operation of the slabs included levelling the surface with a trowel, followed by brushing with a medium-stiff brush after the bleeding of concrete had stopped. After casting, all the molded specimens were covered with water-saturated burlap, and left in the casting room at 24 ± 1.3°C for 24 h. They were then demoulded, and transferred to the moist-curing room until required for testing.

2.4 Testing of specimens

2.4.1 Part A
Two 152x305 mm cylinders from each mixture were tested in compression at both 14 and 28 days. At 14 days, two prisms from each mixture were removed from the

moist-curing room and tested in flexure. Sawn sections of these prisms were used for determining the air-void parameters of the hardened concrete. For mixtures 3 to 10 and 13 to 16, the four remaining prisms were used for freezing and thawing studies done in accordance with ASTM C 666 Procedure A.

The two slabs cast from each mixture were removed from the moist-curing room at the age of 14 days, and were stored in air at a relative humidity of 50 $\pm$ 5% and at a temperature of 23 $\pm$ 1.7°C for an additional period of 14 days. Following this, they were subjected to freezing and thawing cycling in the presence of four per cent calcium chloride solution in accordance with ASTM C 672. The surfaces of the slabs were flushed off at the end of every five cycles and were evaluated visually; the surfaces of the slabs were then covered with a fresh solution of four per cent calcium chloride. The slabs were subjected to a total of 50 cycles of freezing and thawing at a rate of one cycle per day.

2.4.2 Part B

All four 102x203-mm cylinders cast from each mixture were tested in compression at 28 days. Two prisms were tested in flexure at 14 days, and sawn sections from these prisms were used for the determination of the air-void parameters. Eight slabs were cast from each mixture. Table 5 summarizes the types of preparation and testing performed on the slabs in Part B.

Table 5. Type of preparation and type of testing made on slabs of Part B

Slab No.	Type of preparation	Type of testing
1 and 2	14 days moist cured and 14 days air dried	Surface of the slabs covered with a 4% CaCl$_2$ solution, and then subjected to freezing and thawing
3 and 4	28 days moist cured	Same as above
5 and 6	14 days moist cured and 13 days air dried followed by one day in oven at 100°C	Same as above
7	14 days moist cured and 14 days air dried and then the surface was brushed	Same as above
8	14 days moist cured and 14 days air dried	Freezing and thawing with surface of the slab covered with water alone

Slabs 1 and 2 were tested according to ASTM C 672; this required 14 days of moist curing followed by 14 days of air drying before subjecting them to freezing and thawing cycling in the presence of a four per cent calcium chloride solution. Slabs 3 and 4 were moist cured for 28 days, and they were then immediately subjected to the scaling test without being allowed to dry. After 14 days of moist curing and 13 days of air drying, slabs 5 and 6, were placed in an oven and dried at 100°C for 24 hours before the scaling test. Slab 7 was tested according to ASTM C 672 except that the surface of the slab 7 was brushed vigorously with a steel brush to remove a thin layer of mortar before being subjected to the deicing salt scaling test. Slab 8 was cured in accordance with ASTM C 672, but was subjected to freezing and thawing in water alone.

The solutions covering the slabs were changed after every five cycles and the slabs surfaces were evaluated visually. The residue material from scaling was collected and weighed. The slabs were subjected to a total of 50 cycles of freezing and thawing.

2.5 Discussion

2.5.1 Compressive and flexural strengths

The compressive and flexural strength data are given in Table 6. The 28-day compressive strengths ranged from 17.3 to 38.4 MPa and the flexural strengths ranged from 3.2 to 5.8 MPa at 14 days.

Regardless of the W/(C+SF), the compressive strengths of the concretes incorporating silica fume are consistently higher than that of the reference concretes without silica fume. A similar pattern is observed for the flexural strengths.

Table 6. Summary of compressive and flexural strengths

Part	Mix No.	W/(C+SF)	Silica fume, (%)	Compressive strength, (Mpa)		Flexural strength, (MPa)
				14 days	28 days	14 days
	1	0.40	0	28.9	30.7	5.5
	2	0.40	8	33.8	38.4	5.8
	3	0.45	0	26.5	29.9	5.2
	4	0.45	8	27.3	32.8	5.4
	5	0.50	0	25.9	28.3	4.1
	6	0.50	8	28.8	33.1	4.5
	7	0.55	0	21.2	23.5	3.2
	8	0.55	8	25.4	30.1	4.2
A	9	0.60	0	19.7	21.8	3.6
	10	0.60	8	22.8	26.0	3.8
	11	0.65	0	17.2	18.5	3.6
	12	0.65	8	21.5	24.8	3.8
	13	0.55	0	20.6	22.7	4.0
	14	0.55	8	24.8	26.9	4.6
	15	0.60	0	19.2	20.4	4.0
	16	0.60	8	21.9	24.3	4.1
	17	0.65	0	15.9	17.3	3.3
	18	0.65	8	20.3	22.0	3.7
B	19	0.60	0	----	21.7	4.2
	20	0.60	8	----	27.0	4.4

On the other hand, the concrete made using a superplasticizer has a slightly higher compressive strength than the corresponding concrete made without the super-plasticizer, and this is in spite of the fact that the latter type of concrete has a cement content higher than that of the former. Such trend, however, is not reflected in the 14-day flexural strength test results.

2.5.2 Resistance of concrete prisms to freezing and thawing cycling
The test data indicate that all test prisms performed excellently when subjected to repeated cycles of freezing and thawing, with durability factors ranging from 96 to 100 (Table 7). The data on the air-void parameters of the hardened concrete (Table 8) show low spacing factors which explain the good performance of concrete prisms to freezing and thawing cycling.

Table 7. Durability factors after 300 cycles of freezing and thawing

Mixture No.	W/(C+SF)	Silica fume content, (%)	Durability factor
3	0.45	0	98
4	0.45	8	99
5	0.50	0	100
6	0.50	8	99
7	0.55	0	99
8	0.55	8	98
9	0.60	0	96
10	0.60	8	97
13	0.55	0	97
14	0.55	8	98
15	0.60	0	99
16	0.60	8	97

Table 8. Air-void parameters of hardened concretes

Mix No.	W/(C+SF)	Silica fume content, (%)	Air content, (%)	Specific surface, (mm^{-1})	Spacing factor, (μm)
1	0.40	0	6.1	18.6	200
2	0.40	8	6.8	28.5	112
3	0.45	0	5.6	21.1	182
4	0.45	8	8.4	21.6	101
5	0.50	0	6.4	20.0	145
6	0.50	8	7.5	21.0	115
7	0.55	0	7.3	23.1	109
8	0.55	8	9.5	20.3	102
9	0.60	0	8.9	13.6	140
10	0.60	8	7.6	19.8	108
11	0.65	0	8.0	13.9	143
12	0.65	8	7.7	19.0	95
13	0.55	0	6.5	23.4	114
14	0.55	8	6.8	32.2	87
15	0.60	0	7.7	20.8	118
16	0.60	8	6.6	29.2	102
17	0.65	0	8.6	11.7	159
18	0.65	8	5.9	28.6	102
19	0.60	0	5.6	27.7	143
20	0.60	8	5.5	30.9	118

2.5.3 Deicing salt scaling resistance of concrete slabs

The visual evaluation of the surface condition of the slabs of Part A after 50 cycles is shown in Table 9. In general, the data show that all the slabs of Part A performed reasonably well under the scaling test, the worst surface condition observed for any of the slabs after completion of 50 freeze-thaw cycles corresponded to a rating "better than moderate" (3-). Again, in general, the good performance of the concretes in the scaling test even at fairly high W/(C+SF) is certainly due to the adequate air-void spacing factors of the concretes.

It can be noted that the surface deterioration of the concrete tends to increase with an increase in the W/(C+SF), however, such increase remains mostly marginal at W/(C+SF) up to at least 0.55. On the other hand, the concrete incorporating silica fume clearly shows slightly more scaling than the reference concrete, regardless of the W/(C+SF). However, it is doubtful if the observed differences are of any practical significance in view of the very low level of surface deterioration observed in both cases.

As for the effect of the superplasticizer, the results indicate that there is no noticeable difference in the deterioration between the concretes made with and without the superplasticizer.

Table 9. Visual rating of the concrete surfaces after 50 cycles of freezing and thawing: Part A

Mix No.	W/(C+SF)	Silica fume content, (%)	Visual rating*
1	0.40	0	1-
2	0.40	8	1
3	0.45	0	1-
4	0.45	8	1
5	0.50	0	1
6	0.50	8	2-
7	0.55	0	1
8	0.55	8	2-
9	0.60	0	1
10	0.60	8	2
11	0.65	0	1+
12	0.65	8	3-
13	0.55	0	1
14	0.55	8	2-
15	0.60	0	1
16	0.60	8	2+
17	0.65	0	1+
18	0.65	8	2

* 0 - No scaling
  1 - Very slight scaling
  2 - Slight to moderate scaling
  3 - Moderate scaling
  4 - Moderate to severe scaling
  5 - Severe scaling

The results of the final evaluation of the slabs of Part B are given in Table 10. Slabs 1 and 2 of each mixture were prepared and tested as in Part A, i.e., according to ASTM C 672 test procedures. As can be seen in Table 10, both slabs cast from the silica fume mixture again show slightly more surface scaling than those cast from the reference concrete.

As expected, the slabs that performed best were those subjected to freezing and thawing cycling in water only (slab 8). For each mixture, these slabs show very little surface deterioration as compared with slabs 1 and 2, which were prepared and cured the same way, but tested in the calcium chloride solution. This result confirms the primary role of the salt in the scaling process of the concrete subjected to freezing and thawing, regardless of the presence of silica fume in the concrete.

The slabs that had their surfaces brushed before testing also had relatively good resistance to scaling, and this applies for both types of mixtures. The brushing action removed a thin layer of mortar from the surface of the slab. The exposed new surface performed particularly well in the test. These results tend to demonstrate that upon normal finishing and pre-conditioning, there exists a thin layer of mortar at the surface of the slab which is particularly susceptible to the action of de-icing salts under freezing and thawing. This susceptibility of the surface of the slabs to scale is possibly due to the deterioration of the air-void system in this thin layer of mortar during the casting or the finishing of the specimens. The air-void parameters of the concrete surface exposed to the scaling test was possibly very different from those determined on the polished sections cut from the concrete prisms.

Table 10. Visual rating of the concrete surfaces after 50 cycles of freezing and thawing: Part B

Mix No.	Slab No.	Conditioning and testing	Visual rating
	1 and 2	ASTM C 672	2−
	3 and 4	28-d moist cured	1
19	5 and 6	Dried in oven	4−
	7	Brushed surface	1−
	8	No salt	0+
	1 and 2	ASTM C 672	2+
	3 and 4	28-day moist cured	1+
20	5 and 6	Dried in oven	3+
	7	Brushed surface	1
	8	No salt	1−

Slabs 3 and 4, which were continuously moist-cured for 28 days before being exposed to freezing and thawing cycling, also performed better than slabs 1 and 2. This result may be attributable to the longer moist curing period, and therefore increased strength, or to the fact that the concrete surface was not allowed to dry before being subjected to freezing and thawing cycling, or both.

Slabs 5 and 6 performed rather poorly under the scaling test. The high drying temperature used in the pre-conditioning of the specimens probably caused microcracking of the concrete and rendered the slabs quite vulnerable to both the combined action of freezing and thawing and the deicing salts.

In general, the visual evaluation of the slabs of Part B shows that, under all conditions, the reference concrete performed better than the silica fume concrete when subjected to the scaling test, except for the slabs dried in the oven. Nevertheless for all slabs the difference between the two types of concrete is marginal. It appears from the above results that the performance of the concrete in the scaling test is related to some characteristics of the surface of the specimens. These would include the air-void parameters, the strength, and the presence of microcracks.

Figure 1 shows the variation in the mass of the scaling residue as a function of the number of freezing and thawing cycles for the different slabs from the two mixtures of Part B. The graphs confirm the results shown in Table 10 fairly well. Meanwhile, there is no exact relation between the rating and the determined quantities of scaling residue. Figure 1 shows that for slabs 1 and 2 it is principally during the first five cycles that the silica fume concrete deteriorates more than the reference concrete; after that, both concretes deteriorate almost at the same rate. A similar observation can be made for slab 7, but in this case the quantities of scaling residue are smaller. These results indicate that the difference in the scaling resistance between the silica fume concrete and the reference concrete exists mainly at the very top surface of the concrete specimens.

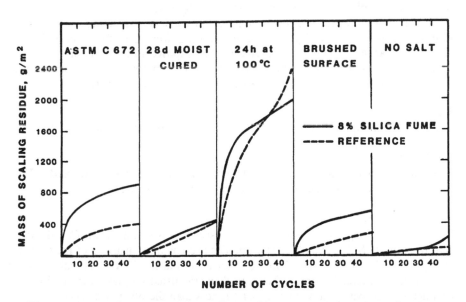

Fig. 1. Cumulative mass of the scaling residue versus the number of cycles

Slabs 3 and 4 deteriorated almost equally for both types of concrete. The oven-dried slabs 5 and 6 made with silica fume concrete scaled faster during the first five cycles than those made with the reference concrete, but subsequently the latter deteriorated faster and eventually produced more scaling residue than the former.

Slabs of silica fume concrete and reference concrete, when subjected to the scaling test with water alone deteriorated at almost the same rate for the major part of the test, but during the last cycles the silica fume concrete showed more scaling.

2.6 Conclusions

In general, all concretes investigated performed satisfactorily under the action of deicing salts. Although it appears that concrete incorporating silica fume is very marginally more susceptible to scaling than concrete without silica fume, the general level of scaling remains low for concretes with W/(C+SF) below 0.60, and the difference between the two types of concrete is then more or less negligible. It is also possible that this difference may be due to the particular conditions and procedures used in the test. The results of preliminary investigations in this regard clearly indicate that the method of preparation and curing of the test specimens has a significant influence on the deicing salt scaling resistance of the concrete.

3 Deicing salt scaling resistance of blast-furnace slag concrete

In this investigation, eight air-entrained concrete mixtures were made with each mixture having a water-to-(cement+slag) ratio (W/(C+S)) of 0.55. Three different slags, two from Canada and one from the U.S.A., were used as a partial replacement for cement at replacement levels of 25 and 50% by mass.

From each concrete mixture, a number of cylinders and prisms were cast for testing in compression and flexure, and test slabs were cast for determining the scaling resistance of the concrete surfaces when exposed to the combined action of freezing and thawing cycles and deicing salts. Sawn sections of the test prisms were used for determining the air-void parameters of the hardened concrete.

3.1 Materials
The concrete mixtures were prepared using the following materials.

3.1.1 Cement
Two different brands of normal ASTM Type I portland cements were used. Their physical properties and chemical analyses are given in Table 11.

3.1.2 Blast-furnace slag
Three different slags were used. These consisted of two slags from Canada and a slag from the U.S.A. The physical properties and chemical analyses of the slags are also given in Table 11. The Canadian pelletized and the U.S.A. granulated slags were from commercial sources. The Canadian granulated slag was from a pilot plant production, and was ground to a higher fineness than the other two slags.

3.1.3 Aggregates
The coarse and fine aggregates consisted of minus 19 mm crushed dolomitic limestone and natural sand, respectively. The grading and physical properties of the coarse and fine aggregates are given in Tables 2 and 3.

3.1.4 Air-entraining admixture

A sulfonated hydrocarbon type air-entraining admixture was used in all the mixtures.

Table 11. Physical properties and chemical analysis of the cements and the slags

	Cements		Blast-furnace slags		
	Brand A	Brand B	Canadian granulated	Canadian pelletized	U.S.A granulated
Physical properties					
Time of set, (min.)					
initial:	130	130	---	---	---
final:	235	---	---	---	---
Fineness;					
45 μm (passing), (%)	82.7	92.8	89.8	83.0	99.0
surface area;					
Blaine, (m^2/Kg)	366	---	608	418	540
Soundness;					
autoclave expansion, (%)	0.13	0.16	0.13	0.04	0.02
Compressive strength of 51-mm cubes, (MPa)					
3 d	22.6	25.1	---	---	---
7 d	27.0	31.5	---	---	---
28 d	32.9	39.3	---	---	---
Chemical analysis, (%)					
SiO_2	20.14	21.24	38.00	38.80	35.10
Al_2O_3	5.40	4.09	8.74	6.55	9.34
Fe_2O_3	2.38	2.81	0.55	1.30	0.98
CaO	62.04	62.21	32.00	35.10	40.10
MgO	2.62	3.75	18.60	12.10	9.62
Na_2O	0.24	-	0.22	0.37	0.18
K_2O	1.12	-	0.76	0.47	0.22
SO_3	3.87	3.03	2.45	3.30	2.90
Bogue compound composition, (%)					
C_3S	48.7	51.7	---	---	---
C_2S	21.0	21.9	---	---	---
C_3A	10.3	6.1	---	---	---
C_4AF	7.2	8.6	---	---	---

3.2 Mixture proportions

The proportioning of the concrete mixtures and the properties of the fresh concrete are summarized in Table 12. For all the mixtures, the graded coarse and fine aggregates were used in the saturated condition.

A total of eight concrete mixtures having a W/(C+S) of 0.55 were made. Four mixtures were made using ASTM Type I cement of Brand A. These consisted of a reference mixture and three mixtures incorporating 50% slag as a partial replacement for cement by mass. The remaining four mixtures were made using ASTM Type I cement of Brand B, and included one reference mixture and three mixtures incorporating 25% slag as a partial replacement for cement.

As far as possible, the slump of the concrete mixtures was maintained at 75 ± 25 mm, and the air content was kept at 6.5 ± 0.5%. The mixing was carried out in a laboratory counter-current pan mixer with the simultaneous addition of the slag and the cement.

Table 12. Mixture proportions and properties of fresh concrete

Mix No.	W/(C+S)	Cement, Brand	(kg/m³)	Slag, (kg/m³)	AEA,* (mL/m³)	Properties of fresh concrete Slump, (mm)	Air, (%)
1	0.55		268	0	160	45	6.5
2	0.55	A	133	133[1]	233	45	6.2
3	0.55		132	132[2]	249	45	6.7
4	0.55		132	132[3]	366	40	6.7
5	0.55		271	0	161	65	6.2
6	0.55	B	206	69[1]	288	90	6.1
7	0.55		203	68[2]	204	70	6.4
8	0.55		200	67[3]	201	100	6.9

* Air-entraining admixture
(1) Canadian granulated slag
(2) Canadian pelletized slag
(3) U.S. granulated slag

3.3 Preparation and casting of test specimens
Two 152x305 mm cylinders, two 76x102x406 mm prisms, and two 280x300x75 mm slabs were cast from each mixture. All cylinders were cast in two layers, each layer being compacted using an internal vibrator. The prisms and the slabs were also cast in two layers, each layer being compacted using a vibrating table. The finishing operation of the slabs included levelling the surface with a trowel, followed by brushing with a medium-stiff brush after the bleeding of concrete had stopped.

After casting, all the moulded specimens were covered with water-saturated burlap and left in the casting room at 24 ± 1.3°C for 24 h. They were then demoulded, and transferred to the moist-curing room until required for testing.

3.4 Testing of specimens
From each mixture, two 152x305-mm cylinders were capped and tested in compression at 28 days, and two prisms were tested in flexure at 14 days. Sawn sections from these prisms were used for determining the air-void parameters of the hardened concrete.

The two slabs cast from each mixture were removed from the moist-curing room at the age of 14 days. These were then stored in air at a relative humidity of 50 ± 5% and at a temperature of 23 ± 1.7°C for an additional period of 14 days for the mixtures 1 to 4, and for a period of 49 days for the mixture 5 to 8. This longer period of air curing for the slabs of concrete mixtures 5 to 8 was due to the break down of the freezing equipment.

After the air-curing period, the slabs were subjected to the freezing and thawing cycling in the presence of a five per cent sodium chloride solution. The surfaces of the slabs were flushed off at the end of every five cycles, and they were then evaluated visually. Following this, the slabs were covered with a fresh solution of five per cent sodium chloride. The residue material from the scaling was collected and weighed. The slabs were subjected to a total of 50 cycles of freezing and thawing at a rate of one cycle per day.

3.5 Discussion

3.5.1 Compressive and flexural strengths
The compressive and flexural strength data are given in Table 13. The 28-day compressive strengths ranged from 23.3 to 30.4 MPa; the 14-day flexural strengths ranged from 3.5 to 4.8 MPa.

For both percentages of cement replacement by slag, the compressive and flexural strengths of the concrete incorporating slag are comparable with the strength of the reference concrete. The only exception is the mixture incorporating 25% of the granulated slag from Canada (Blaine 608 m^2/kg); the test specimens from this mixture show a significantly higher strength than that of the reference concrete.

Table 13. Compressive and flexural strengths of concrete

Mix No.	Cement brand	W/(C+S)	Slag, (%)	Compressive strength at 28 days, (MPa)	Flexural strength at 14 days, (MPa)
1		0.55	0	25.9	4.5
2	A	0.55	50[1]	26.8	4.5
3		0.55	50[2]	23.3	3.6
4		0.55	50[3]	26.4	4.5
5		0.55	0	26.7	3.5
6	B	0.55	25[1]	30.4	4.8
7		0.55	25[2]	25.1	3.8
8		0.55	25[3]	26.1	3.9

(1) Canadian granulated slag
(2) Canadian pelletized slag
(3) U.S. granulated slag

3.5.2 Deicing salt scaling resistance of concrete slabs
The visual evaluation of the surface condition of the slabs after 50 cycles of freezing and thawing is shown in Table 14. The reference concrete performed rather well under the scaling test. The surface condition observed for the slabs after the completion of 50 cycles of freezing and thawing corresponded to a rating of "very slight scaling". The reference concrete made with portland cement Brand A performed slightly better than the one made with portland cement Brand B, the ratings being 1- for the former and 1+ for the latter.

On the other hand, the test data show that the concrete incorporating the slags suffered more deterioration than the reference concretes. This is true at both levels of cement replacement by the slags. In general, the surface condition of all the slabs after the scaling test corresponded to a scale rating of "moderate".

The type of slag used appears to have little noticeable influence on the degree of deterioration of the concrete surfaces. However, the test data show that concrete incorporating 25% slag performed somewhat better than the concrete incorporating 50% slag, indicating the effect of the percentage of the slag used. Concerning this last observation, it should be taken into account that the concretes with 25% slag were made using cement Brand B, and the slabs had been subjected to a longer air-curing period. These might have influenced the scaling test results.

Table 14. Visual rating of the concrete surfaces after 50 cycles of freezing and thawing and air-void parameters of the hardened concrete

Mix No.	Cement brand	W/(C+S)	Slag (%)	Visual rating	Air-void parameters	
					Air content, (%)	Spacing factor, (μm)
1		0.55	0	1-	8.2	124
2	A	0.55	50	3+	5.1	130
3		0.55	50	3+	8.4	141
4		0.55	50	3+	7.3	101
5		0.55	0	1+	7.0	114
6	B	0.55	25	3	9.1	102
7		0.55	25	3	9.2	098
8		0.55	25	3	---	---

The somewhat poor performance of the slag concretes in the scaling test may be partly due to the fairly high W/(C+S) of the concretes investigated. However, the reference concretes, which showed much better scaling resistance, had the same W/(C+S), and the compressive strengths at 28 days and the flexural strengths at 14 days of reference and slag concretes were very similar.

Figure 2 shows the variation in the mass of the scaling residue as a function of the number of freezing and thawing cycles. This confirms the results of the visual evaluation shown in Table 14. The slag concretes show much higher amounts of the scaling residue than the reference concrete. Furthermore, the mass of the scaling residue for the concrete incorporating 50% slag was slightly more than that of the concrete incorporating 25% slag.

It is also observed that the rate of surface deterioration was higher during the first 10 cycles for the concretes investigated, especially for the slag concretes. In fact, a large proportion of the difference in the total amount of scaling residue between the slag concretes and the reference concretes is the result of the very fast rate of deterioration of the slag concrete in the first 10 cycles of freezing and thawing. This suggests that there was at the surface of the slag concrete specimens a layer of concrete much more susceptible to scaling than the rest of concrete deeper in the slab specimen, or than the reference concrete, even at its top surface.

The data on air-void parameters presented in Table 14 show low spacing factors for all concretes. The poor performance of the slag concretes compared to that of reference concretes under the action of de-icing salts cannot be attributed to the lack of adequate air-void system in the hardened concrete. However, as it was mentioned in the above silica fume concrete study, it should be noted that the air-void parameters were determined on specimens cut from prisms and may be different from those at the surface of the slabs.

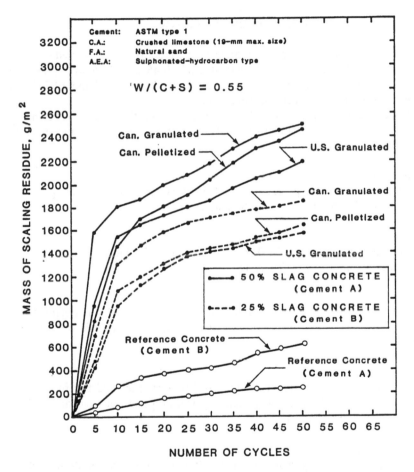

Fig. 2. Mass of scaling residue versus number of cycles for slag concretes and reference concretes (from Ref. 14)

3.5 Conclusions

Regardless of the percentage and the type of the slag used, the air-entrained concrete incorporating slag exhibited considerably more surface scaling than the reference concrete. According to the ASTM visual rating scale, the scaling of the concrete slabs incorporating slag was classified as "moderate". The mass of the scaling residue was in excess of 1400 g/m².

The surface of the concrete slabs incorporating 50% slag as replacement for cement showed somewhat more scaling than the surface of the concrete slabs incorporating 25% slag.

The test results suggest that the somewhat poor scaling resistance of slag concrete compared to that of reference concrete is mainly due to the thin layer of mortar at the surface of the slag concrete slab specimen, which is more susceptible to scaling.

4   Deicing salt scaling resistance of fly ash concrete

In this investigation, a total of 21 air-entrained concrete mixtures were made. Water-to-(cement+fly ash) ratios (W/(C+FA)) of 0.35, 0.45 and 0.55 were used, and reference concrete (without fly ash) and concrete incorporating 20 and 30% fly ash as replacement by mass for cement were made.

Test specimens were cast for the determination of the scaling resistance of concrete, and various combinations of moist curing, membrane curing and air drying were included in the study.

4.1 Materials
The concrete mixtures were prepared in the laboratory in a counter current pan mixer of 0.15 m³ capacity using the following materials:

4.1.1 Cement
Normal portland cement, ASTM Type I, was used. Its physical properties and chemical analysis are given in Table 15.

4.1.2 Fly ash
The fly ash used was a low-calcium type (ASTM Class F), and was obtained from a source in Nova Scotia. Its physical properties and chemical analysis are also given in Table 15.

4.1.3 Aggregates
The coarse fraction of the aggregate was crushed quartzite, and the fine fraction consisted of natural sand. The grading and physical properties of the coarse and fine aggregates are given in Tables 16 and 17.

4.1.4 Air-entraining admixture
A neutralized vinsol resin type air-entraining admixture was used for all the concrete mixtures.

4.1.5 Water-reducing agent
A lignin-based water reducer with 33% total solids was used as a water-reducing agent.

Table 15. Physical properties and chemical analysis of the cement and the fly ash

	Portland cement (ASTM Type I)	Fly ash (ASTM Class F)
Physical properties		
Fineness;		
45 μm (passing), (%)	94.3	84.0
surface area; Blaine, (m^2/kg)	368	---
Specific gravity;	---	2.54
Setting time; (min.)		
initial:	140	---
final:	280	---
Compressive strength of 51-mm cubes, (MPa)		
3 d	28.8	---
7 d	33.3	---
28 d	38.7	---
Chemical analysis, (%)		
SiO$_2$	21.26	45.20
Al$_2$O$_3$	4.48	20.70
Fe$_2$O$_3$	1.71	24.83
CaO	62.88	1.63
MgO	3.23	0.97
K$_2$O	0.49	2.40
Na$_2$O	0.50	0.59
SO$_3$	2.35	---
TiO$_2$	----	1.05
Loss on ignition	2.17	2.66
Bogue compound composition, (%)		
C$_3$S	55.2	---
C$_2$S	19.3	---
C$_3$A	9.0	---
C$_4$AF	5.2	---
Pozzolanic Activity Index, (ASTM C 311), (%)	---	96.0

Table 16. Grading of aggregates

Coarse aggregate		Fine aggregate	
Sieve size, (mm)	Percentage passing	Sieve size, (mm)	Percentage passing
20	100	10	100
14	79.3	5	98
10	43.5	2.5	89
5	8.4	1.25	70
2.5	2.5	0.630	43
		0.315	18
		0.160	5

Table 17. Physical properties of aggregates

	Coarse aggregate	Fine aggregate
Specific gravity (SSD)	2.71	2.61
Absorption, (%)	0.39	1.3

4.2 Mixture proportions

The proportioning of the concrete mixtures is summarized in Table 18. The mixtures in series I were proportioned with water-to-(cement+fly ash) ratios (W/(C+FA)) of 0.35, 0.45 and 0.55, whereas the W/(C+FA) was kept constant at 0.45 for the mixtures of series II. In each series and for each W/(C+FA), a reference concrete (without fly ash) and concretes incorporating 20 and 30% fly ash as replacement by mass for cement were made. An air content of 6.5 ± 0.5% was specified for all the mixtures. The properties of the fresh concrete are shown in Table 19.

Table 18. Mixture proportions

Mix Series	Mix No.	W/(C+FA)	Cement, (kg/m³)	Fly ash, (%)	Fly ash, (kg/m³)	AEA,* (mL/m³)	WR,** (mL/m³)
	1	0.35	458	0	0	105	2130
	2	0.45	353	0	0	65	1645
	3	0.55	289	0	0	55	1325
	4	0.35	368	20	92	130	1975
I	5	0.45	288	20	72	130	1645
	6	0.55	234	20	59	130	1325
	7	0.35	324	30	139	205	2120
	8	0.45	249	30	107	160	1645
	9	0.55	203	30	87	150	1325
	10	0.45	357	0	0	195	1650
II	11	0.45	283	20	71	205	1645
	12	0.45	248	30	106	195	1645

* Air-entraining admixture
** Water-reducer

4.3 Preparation and casting of test specimens

Twenty four 152x305x75 mm slabs were cast from each mixture and were compacted by rodding. The slabs were used for the determination of the deicing salt scaling resistance of concrete in accordance with ASTM standard C 672.

After casting, the slab specimens were covered with water saturated burlap, and were left in the casting room at 24 ± 1.3°C and 50% relative humidity for 24 h. All specimens were then demoulded. The slabs from mixtures of series I were transferred to the moist-curing room at 100% relative humidity until required for testing. Slabs from mixtures of series II were covered with a curing compound, and left in the laboratory until required for testing. Additional batches of concrete were made for selected mixtures of series I to provide concrete for casting 150x300 mm cylinders for the compressive strength determination.

Table 19. Properties of fresh concrete

Mix Series	Mix No.	W/(C+FA)	Cement (kg/m³)	Fly ash (%)	(kg/m³)	Slump, (mm)	Air (%)
	1	0.35	458	0	0	90	7.0
	2	0.45	353	0	0	90	6.2
	3	0.55	289	0	0	70	6.0
	4	0.35	368	20	92	65	5.9
I	5	0.45	288	.20	72	110	6.0
	6	0.55	234	20	59	110	6.0
	7	0.35	324	30	139	120	7.0
	8	0.45	249	30	107	85	6.6
	9	0.55	203	30	87	100	6.6
	10	0.45	357	0	0	45	6.2
II	11	0.45	283	20	71	75	6.5
	12	0.45	248	30	106	90	6.4

4.4 Testing of specimens

The 150x300 mm cylinders were capped and tested in compression at different ages up to one year.

The schedule of conditioning of the test slabs for the deicing salt scaling tests is shown in Table 20. At the end of each of the three specified curing periods of 3, 7 and 14 days, eight concrete slabs from each mixture of series I were removed from the moist-curing room and stored in air for different periods of time. Two of these eight slabs were air dried for three weeks, two for four weeks, two for five weeks and two for six weeks.

The slabs were then subjected to freezing and thawing in the presence of a four per cent calcium chloride solution in accordance with ASTM C 672. The salt solutions were changed every five cycles, and the slab surfaces were evaluated visually. The residue material from scaling was collected and weighed. The slabs were subjected to a total of 50 cycles of freezing and thawing at the rate of one cycle per day.

Table 20. Schedule of conditioning of slab specimens

Curing	Drying			
	3 weeks	4 weeks	5 weeks	6 weeks
3 days	2 slabs	2 slabs	2 slabs	2 slabs
7 days	2 slabs	2 slabs	2 slabs	2 slabs
14 days	2 slabs	2 slabs	2 slabs	2 slabs

At the end of the three specified curing periods, eight concrete slabs from each mixture of series II had their membrane removed on the surface by brushing with a steel brush. After brushing, the slabs were left to dry in air for the same periods of time as the moist-cured slabs of series I. The slabs were then subjected to the scaling test following the procedures outlined above.

The air-void parameters of the hardened concrete were determined for each mixture of each series using cut sections of test slabs after completion of the scaling tests.

4.5 Discussion

4.5.1 Compressive strength
The non-fly ash reference concretes show consistently higher compressive strengths than the concretes incorporating 30% fly ash up to the age of 28 days (Table 21). The strengths for the reference concretes and fly ash concretes are generally comparable at 91 and 365 days. The actual strength of the concrete at the age of the scaling test was not determined and is not known precisely. Unlike the cylinder specimens used for strength test, the slab specimens were moist or membrane cured for 3, 7 or 14 days and air dried for different periods of time before they were tested for scaling resistance. It is to be expected that some gain in strength occurred during the drying period. However, from the strength data available, it can be reasonably estimated that in each case the compressive strength of the reference concrete was probably somewhat higher than that of the concrete incorporating 20 or 30% fly ash at the age of testing for scaling resistance.

Table 21. Mechanical properties of selected concrete mixtures from series I

Mix No.	W/(C+FA)	Cement, (kg/m³)	Fly ash, (%)	Fly ash, (kg/m³)	Compressive strength, (MPa)				
					3 d	7 d	28 d	91 d	365 d
1	0.35	457	0	0	36	41	47	51	54
2	0.45	353	0	0	27	32	36	40	42
3	0.55	287	0	0	21	26	33	34	38
7	0.35	320	30	137	24	31	39	50	58
8	0.45	247	30	106	16	20	30	42	43
9	0.55	201	30	86	10	15	23	31	37

4.5.2 Air-void parameters
The values of the air-void spacing factor of the hardened concretes shown in Table 22 are all of the order of 200 μm which is the recommended limit to ensure good resistance to freezing and thawing cycling [17].

4.5.3 Deicing salt scaling resistance
The final visual evaluation of the surface condition of the slabs after 50 cycles of freezing and thawing is given in Tables 23 and 24.

The data show that independently of the concrete composition and the curing conditions, the slabs performed reasonably well under the scaling test when evaluated visually; the large majority of the slabs were rated 2 or 2.5, corresponding to a surface condition of slight to moderate scaling according to the ASTM scaling test. The data in Table 23 show that the deterioration of the concrete tends to increase slightly with an increase in the W/(C+FA). It can be noted that the membrane-cured concrete performed slightly better than the corresponding moist-cured concrete, as evident from the comparison between the mixtures of series I and series II. The data also show that based on the visual assessment, the incorporation of fly ash does not significantly affect the salt scaling resistance of concrete. This is generally true for all concrete tested.

Table 22. Air-void parameters of hardened concrete

Mix Series	Mix No.	W/(C+FA)	Fly ash, (%)	Air content, (%)	Specific surface, (mm^{-1})	Spacing factor, (μm)
	1	0.35	0	8.9	17.0	216
	2	0.45	0	7.6	18.8	202
	3	0.55	0	8.2	16.6	202
	4	0.35	20	6.6	22.4	192
I	5	0.45	20	8.7	18.1	201
	6	0.55	20	7.3	18.4	210
	7	0.35	30	6.0	25.4	176
	8	0.45	30	6.3	22.4	185
	9	0.55	30	5.2	20.2	220
	10	0.45	0	7.5	19.0	211
II	11	0.45	20	8.9	18.9	144
	12	0.45	30	7.0	21.2	178

Table 23. Final visual evaluation of the concrete slabs of mixture series I after 50 cycles of freezing and thawing

		Final visual rating								
		W/(C+FA) = 0.35			W/(C+FA) = 0.45			W/(C+FA) = 0.55		
Moist curing,	Drying,	Fly ash, (%)			Fly ash, (%)			Fly ash, (%)		
(days)	(weeks)	0	20	30	0	20	30	0	20	30
	3	2.5	3.5	2	3	2	2	2	2.5	2
3	4	2.5	2.5	2	2.5	2	3.5	2	2.5	3.5
	5	2	2	2	2	2.5	3	2	3.5	2
	6	2	2	2	2	2	2	2	3	2
	3	2.5	2	2.5	2	3.5	2.5	2.5	3.5	2.5
7	4	2	2	2.5	2	4	2.5	2.5	2.5	3
	5	2	2	3	2	2	2	2	2	2
	6	2	2	2	2.5	2.5	2.5	2	2.5	2.5
	3	2	3.5	4.5	2.5	2.5	2.5	3	4	2.5
14	4	2	2.5	3.5	2	2	2	2	4.5	2.5
	5	3	2	3	2	3.5	3.5	2.5	4	2.5
	6	2	2	2.5	2.5	2.5	2	3.5	4	3

The mass of the scaling residue at the end of the scaling test is shown in Fig. 3 to 6 for different times of moist or membrane curing and times of air drying. Figures 3 to 6 show the effect of the different factors such as W/(C+FA), percentage of fly ash, time of curing and drying, and the use of membrane curing on the scaling resistance of concrete.

Figures 3 to 5 show that the mass of scaling residue tends to increase with an increase in the W/(C+FA). This increase is somewhat more marked for concretes incorporating fly ash.

In general, for concretes having the same W/(C+FA), and subjected to the same curing and drying conditions, the incorporation of fly ash increases the mass of scaling residue. This fact is more evident for higher W/(C+FA). It is also more visible for the moist-cured concretes than for the membrane-cured concretes. Notwithstanding the above, the mass of the scaling residue is, in general, well within the acceptance limit of 0.80 kg/m² which has been adopted in certain specifications[1].

Table 24.   Final visual evaluation of the concrete slabs of mixture series II after 50 cycles of freezing and thawing

| Membrane curing, (days) | Drying, (weeks) | Final visual rating | | |
| | | Percentage of fly ash | | |
		0	20	30
3	3	2	2	2
	4	2	2	2
	5	1.5	2	2
	6	2	2	2
7	3	2.5	2	2
	4	2	2.5	2.5
	5	1.5	2	2
	6	2	2	2.5
14	3	1	2	2.5
	4	2	2	2
	5	1	2	2
	6	2	2	2

Overall, there is no beneficial effect of longer moist-curing or membrane-curing periods. In fact, in many cases, concretes subjected to longer curing periods showed more scaling. This was noticed mainly for moist-cured fly ash concretes. The general lack of influence of curing duration may actually be due to the fact that the scaling resistance was found to be relatively good in most cases after only 3 days of curing.

The test data also show that there is no specific trend with respect to the effect of the duration of air drying period on the scaling resistance of concrete. However, the data reflect the high variability of the scaling test results, especially for the fly ash concretes.

The mass of scaling residue is reduced noticeably when membrane curing is used instead of moist curing (compare Fig. 4 to Fig. 6). The better resistance of membrane-cured concrete to deicing salt scaling has been shown by other investigators [12, 18]. In this investigation, the beneficial action of the membrane is more marked for fly ash concretes than for reference concretes. The lower mass of scaling residue for the membrane-cured specimens might be partly due to the removal by the

---

[1]Ministry of Transportation of Ontario, Toronto.

brushing action of a very thin layer of concrete particularly susceptible to scaling at the surface of the slabs, similarly to the results presented in the above study on silica fume concrete.

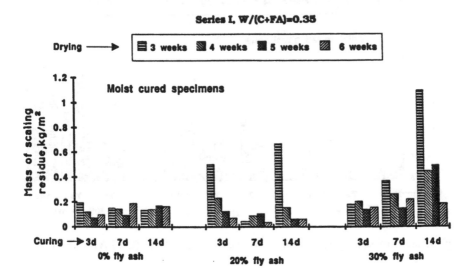

Fig. 3. Mass of scaling residue after 50 cycles: Series I, W/(C+FA)=0.35 (from Ref. 15)

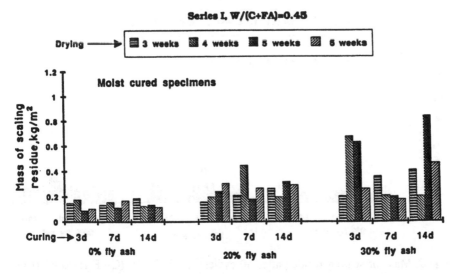

Fig. 4. Mass of scaling residue after 50 cycles: Series I, W/(C+FA)=0.45 (from Ref. 15)

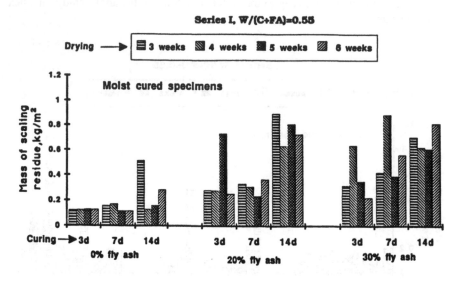

Fig. 5. Mass of scaling residue after 50 cycles: Series I, W/(C+FA)=0.55 (from Ref. 15)

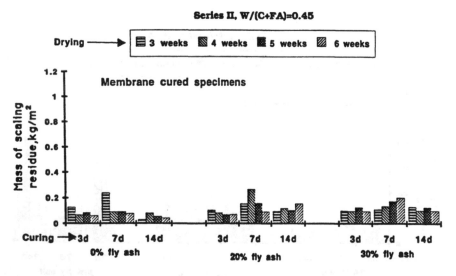

Fig. 6. Mass of scaling residue after 50 cycles: Series II, W/(C+FA)=0.45 (from Ref. 15)

In general, the data show that the concretes performed reasonably well under the scaling test, the mass of scaling residue of the large majority of specimens being well below the above mentioned limit of 0.80 kg/m^2. Nevertheless, the mass of scaling residue of some of the slabs went close and even passed beyond that limit. These slabs are from the moist-cured fly ash concrete.

4.6 Conclusions

According to the visual evaluation or the total mass of scaling residue, the concretes tested generally performed well under the combined action of freezing and thawing in the presence of deicing salts. The large majority of the specimens evaluated exhibited slight to moderate scaling rating or had less scaling residue than the acceptance limit of 0.80 kg/m^2.

Concrete incorporating up to 30% fly ash performed satisfactorily under the scaling test. However, the performance of the fly ash concrete was more variable than the reference concrete. The satisfactory scaling resistance of concrete, even for fly ash mixtures with a W/(C+FA) of 0.55 cured for 3 days is undoubtedly the result of an adequate air-void system in all cases.

Extended moist-curing or drying periods did not affect significantly the performance of the reference and fly ash concretes in the scaling test, at least within the periods investigated.

Membrane curing appears to improve somewhat the durability of concrete under the combined action of freezing and thawing and deicing salts; this is especially so for fly ash concrete.

5 Deicing salt scaling resistance of high-volume fly ash concrete

In this study a total of twelve air-entrained concrete mixtures were made. Reference concretes (without fly ash) and high-volume fly ash concretes (incorporating 58% fly ash as replacement for cement by mass) were made and tested. The water-to-(cement+fly ash) ratio (W/(C+FA)) of the mixtures ranged from 0.27 to 0.39, and fly ash from three different sources were used in the mixtures.

The resistance of concrete to the combined action of freezing and thawing cycling and deicing salts was determined for each concrete mixture.

5.1 Materials
The concrete mixtures were made in the CANMET laboratory using the following materials.

5.1.1 Cement
Normal portland cement, ASTM Type I was used. Its physical properties and chemical analysis are given in Table 25.

5.1.2 Fly ash
The three fly ashes used were of low-calcium variety (ASTM Class F), and were from sources in Nova Scotia (fly ash L), Alberta (fly ash S) and New York State (fly ash R). Their physical properties and chemical analyses are also given in Table 25.

Table 25. Physical properties and chemical analysis of the cement and the fly ashes

	Cement	Fly ash L	Fly ash S	Fly ash R
Physical properties				
Fineness;				
45 μm (passing), (%)	85.3	85.4	85.6	83.7
Blaine, (m^2/kg)	373	318	348	307
Specific gravity;	3.14	2.63	2.04	2.54
Setting time, (min.)				
initial:	165	---	---	---
final:	290	---	---	---
Autoclave expansion, (%)	0.11	---	---	---
Compressive strength of 51-mm cubes, (MPa)				
3 d	18.3	---	---	---
7 d	24.4	---	---	---
28 d	33.3	---	---	---
Chemical analysis, (%)				
SiO_2	21.95	42.2	53.9	48.2
Al_2O_3	4.13	21.6	20.9	24.9
Fe_2O_3	2.88	27.6	3.52	18.9
CaO	61.73	1.87	12.00	2.80
MgO	3.68	0.97	1.11	1.10
SO_3	2.81	1.10	0.09	0.78
Na_2O	0.25	0.66	2.74	0.59
K_2O	0.44	2.55	0.50	1.87
Loss on ignition	1.47	1.85	0.57	3.70
Bogue compound compositions, (%)				
C_3S	42.8	---	---	---
C_2S	30.7	---	---	---
C_3A	6.8	---	---	---
C_4AF	8.8	---	---	---
Pozzolanic activity with portland cement				
Activity index at 28 days, (%)	---	97	92	86

5.1.3 Aggregates

The fine and coarse aggregates were local natural sand and minus 19-mm crushed limestone, respectively. The grading and the physical properties of both aggregates are given in Tables 2 and 3.

5.1.4 Superplasticizer

A sulphonated naphthalene formaldehyde condensate was used.

5.1.5 Air-entraining admixture
A synthetic resin type air-entraining admixture was used in all the mixtures.

5.2 Mixture proportions
The proportioning of the concrete mixtures and the properties of fresh concrete are summarized in Table 26. For all the mixtures, the graded coarse and fine aggregates were used in the water-saturated condition.

The air-entrained, superplasticized concrete mixtures were proportioned to have three different water-to-cementitious materials ratios of 0.39, 0.31, and 0.27. For each $W/(C+FA)$, one reference mixture (without fly ash) and three high-volume fly ash mixtures, each incorporating a different fly ash were made. The proportion of fly ash to total cementitious materials content was 58%. The 0.07 m^3 concrete batches were made in a laboratory counter-current mixer with fly ash added as a separate ingredient.

Table 26. Mixture proportions and properties of fresh concrete

Mix No.	W/(C+FA)	Cement, (kg/m³)	Fly ash, Source	Fly ash, (kg/m³)	SP,* (L/m³)	AEA,** (mL/m³)	Slump, (mm)	Air, (%)
1	0.38	126	L	173	2.5	185	150	5.8
2	0.31	156	L	215	2.8	285	180	5.1
3	0.27	180	L	248	3.7	410	190	5.3
4	0.39	124	S	172	2.2	145	180	4.4
5	0.31	155	S	214	3.2	300	180	4.5
6	0.27	182	S	251	3.5	355	190	5.2
7	0.39	124	R	170	3.3	290	205	5.9
8	0.31	155	R	213	3.7	385	205	4.9
9	0.27	181	R	250	4.4	490	205	4.2
10	0.39	292	--	0	3.9	155	135	5.5
11	0.31	368	--	0	4.4	260	90	4.6
12	0.27	428	--	0	4.9	300	90	4.8

* Superplasticizer
** Air-entraining admixture

5.3 Preparation and casting of test specimens
From each mixture, ten 152x305 mm cylinders were cast and used for determining the compressive strength of concrete at different ages, two 280x300x75 mm slabs were made and used for the determination of the deicing salts scaling resistance of concrete, and a number of 76x102x390 mm prisms were cast for various testing purposes. Sawn sections of some of these prisms were used for the determination of the air-void parameters of hardened concrete. All specimens were cast in two layers with each layer compacted using an internal vibrator for the cylinders and a vibrating table for all the other specimens. After casting, all the moulded specimens were covered with water-saturated burlap, and left in the casting room at 23 ± 1.7 °C and 50% relative humidity. After 24 h, the specimens were demoulded, weighed and transferred to the moist-curing room until required for testing.

5.4 Testing of specimens

The cylinders were tested in compression at ages up to one year. After 28 days of moist curing, the two 280x300x75 mm slabs from each mixture were removed from the moist-curing room and stored in air at a relative humidity of 50 ± 5% and at a temperature of 23 ± 1.7 °C for an additional period of 42 days. They were then subjected to freezing and thawing cycling in the presence of a three per cent sodium chloride solution. Sawn sections of prisms were used for determining the air-void parameters of the hardened concrete.

5.5 Discussion

5.5.1 Compressive strength

The reference concretes exhibited noticeably higher compressive strengths than the fly ash concretes at all ages, and for each W/(C+FA) (Table 27). In general, concretes made with fly ash S showed higher strengths than the other fly ash concretes for each W/(C+FA), especially at the ages of 7, 28 and 91 days. This is the result of the rapid strength development of these concretes between 1 and 28 days, probably due to the higher calcium content of fly ash S compared to that of fly ash L or R.

Table 27. Compressive strength of concrete at different ages

Mix No.	W/(C+FA)	Cement, (kg/m³)	Fly ash, Source	Fly ash, (kg/m³)	Compressive strength, (MPa) 1 d	7 d	28 d	91 d	365 d
1	0.38	126	L	173	3.9	11.2	18.6	30.1	39.4
2	0.31	156	L	215	6.8	16.0	25.2	37.1	47.1
3	0.27	180	L	248	8.6	20.5	32.0	44.1	51.9
4	0.39	124	S	172	3.2	13.2	27.8	35.9	41.1
5	0.31	155	S	214	6.5	22.0	41.6	46.0	51.4
6	0.27	182	S	251	9.9	26.4	45.2	50.3	56.8
7	0.39	124	R	170	3.9	11.2	19.3	30.9	42.7
8	0.31	155	R	213	7.0	17.0	27.5	41.2	51.4
9	0.27	181	R	250	9.2	20.9	36.8	44.2	55.0
10	0.39	292	--	0	16.4	27.8	34.1	40.3	47.1
11	0.31	368	--	0	27.8	43.3	52.7	66.4	78.9
12	0.27	428	--	0	37.2	51.2	60.2	71.8	83.7

5.5.2 Deicing salt scaling resistance

The final visual evaluation of the slabs subjected to the deicing salts scaling test show that all high-volume fly ash concretes performed poorly in this test (Table 28). The high-volume fly ash concrete slabs were given a visual rating of 5, and according to ASTM Standard C 672, this corresponds to severe scaling. The reference concretes performed better than the fly ash concretes. Test slabs cast from reference concrete mixture 10 (W/(C+FA) = 0.39) were given a visual rating of 3- (moderate scaling) whereas the slabs cast from reference mixture 11 (W/(C+FA) = 0.31) and mixture 12 (W/(C+FA) = 0.27) were given a visual rating of 1+ and 2-, respectively.

Table 28. Deicing salt scaling test results and air-void parameters of hardened concrete

Mix No.	W/(C+FA)	Fly ash source	Scaling test results		Air-void parameters	
			Final visual rating	Total scaling residue, (kg/m^2)	Voids in hardened concrete, (%)	Spacing factor, (μm)
1	0.38	L	5	8.66	4.9	213
2	0.31	L	5	6.24	4.9	167
3	0.27	L	5	5.27	4.5	193
4	0.39	S	5	9.08	4.0	204
5	0.31	S	5	5.80	4.1	188
6	0.27	S	5	4.76	3.6	211
7	0.39	R	5	10.64	5.2	256
8	0.31	R	5	7.95	3.5	235
9	0.27	R	5	8.75	2.3	286
10	0.39	--	3-	2.04	4.3	233
11	0.31	--	1+	0.56	3.5	230
12	0.27	--	2-	0.85	3.1	214

The mass of scaling residue at the end of the scaling test is high for all fly ash concretes ranging from 4.76 to 10.64 kg/m^2 (Table 28). In general, the amount of scaling residue tends to decrease with the reduction of the W/(C+FA) of the concrete. These values are much higher than the acceptance limit of 0.80 kg/m^2 stated above in the investigation on the scaling resistance of fly ash concrete.

In general, a value of air-void spacing factor less than 0.200 mm is recommended for satisfactory resistance to repeated cycles of freezing and thawing. Other CANMET investigations have shown that high-volume fly ash concrete with adequate air-void parameters had satisfactory resistance to repeated freezing and thawing cycling test done in accordance with ASTM C 666 Procedure A [7,9]. In this investigation, the spacing factors of concretes made with fly ash L and S ranged from 0.167 to 0.213 and from 0.188 to 0.211, respectively (Table 28). Concretes made with fly ash R and the reference concretes show spacing factors slightly higher than the recommended limit with values ranging from 0.235 to 0.286 for the former, and from 0.214 to 0.233 for the latter. The slightly high spacing factors of the mixtures might be due to the high dosages of the superplasticizer used in the mixtures. As mentioned by some investigators [19-20], superplasticizers have the tendency to increase the average size of the air bubbles and, therefore may increase the air-void spacing factor in the hardened concrete.

All high-volume fly ash concretes performed poorly in the scaling test, even those having air-void spacing factors complying with the recommended value. Fly ash R concretes, which had higher spacing factors, show the highest amounts of scaling residue, ranging from 7.95 to 10.64 kg/m^2. Though the reference concretes showed a much better resistance to deicing salt scaling than the high-volume fly ash concretes, they also exhibited noticeable amounts of scaling residue. For example, the total amount of scaling residue of mixture 10 is 2.04 kg/m^2; much more than the acceptance limit. Also, mixture 12, with a total amount of 0.85 kg/m² is slightly over the limit specified by the Ontario Ministry of Transportation. The amounts of scaling residue of the reference concretes are surprisingly high considering that these concretes have low W/C and high strengths, particularly mixtures 11 and 12. This may be due, at least in part, to the slightly higher air-void parameters of these concretes.

Once again, as it was mentioned in the previous investigations described in this paper, it should be noted that the air-void parameters of concrete at the top layer of the slabs are not known, and may not be adequate for freezing and thawing durability, particularly in the presence of deicing salts. The casting and finishing process of the slabs, combined with the use of high dosages of the superplasticizer might have damaged the air-void system at the surface of the slabs. If so, this could explain the poor performance of the high-volume fly ash concrete and the moderate performance of the reference concrete in the scaling test, but there might be other unknown factors affecting the test results.

Figure 7 illustrates the variation of the cumulative mass of scaling residue as a function of the number of cycles. This shows that, in general, for most of the high-volume fly ash concretes, the rate of deterioration is faster in the first 15 to 20 cycles than in the cycles from 20 to 50. This can be explained by the fact that, once the layer of mortar at the top surface of the specimens has scaled off, the scaling takes place in the mortar around the coarse aggregates, which leaves less surface for scaling, and therefore less scaling residue. Despite that, for all high-volume fly ash concretes, the rate of scaling was still high after that the top layer of mortar was completely removed. This indicates that the scaling occurred not only at the top surface, but also deeper into the concrete specimen.

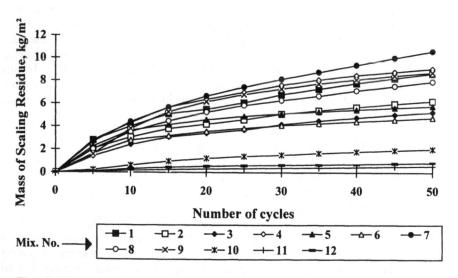

Fig. 7. Mass of scaling residue versus number of cycles for high-volume fly ash concrete

Mixtures 1, 4, and 7 to 9 showed very fast rates of deterioration after 20 cycles. These concretes, particularly mixtures 7 to 9, had also the highest air-void spacing factors of the high-volume fly ash concretes, and this might be one explanation for the very poor performance of these concretes in the scaling test.

5.6 Conclusions

The data presented above show that in spite of the adequate strengths and in some cases low values of the air-void spacing factor, all high-volume fly ash concretes investigated performed poorly when subjected to deicing salt scaling test. The performance was worse for high-volume fly ash concretes having values of the spacing factor slightly higher than that recommended for good resistance to freezing and thawing cycling.

In spite of their high strengths, the reference concretes showed some signs of deterioration and noticeable amounts of the scaling residue.

The poor performance of the high-volume fly ash concretes and the moderate performance of the reference concretes seem to be related to the unsatisfactory quality of the air-void system, which was possibly affected by the use of high dosages of the superplasticizer, and this possibly was aggravated by the finishing process at the surface of the slabs.

6 General conclusions

In general, the incorporation of supplementary cementing materials in concrete affected adversely its deicing salt scaling resistance. This effect ranged from very marginal to considerable depending upon the type and percentage of the supplementary cementing material used.

The concretes incorporating 8% silica fume or up to 30% fly ash performed satisfactorily in the scaling test, and showed only slightly more scaling than the reference concretes of the same water-to-cementitious materials ratio. According to the visual rating scale, the scaling of the surface of concrete incorporating silica fume was classified as slight or slight to moderate, while it was classified as slight to moderate for the large majority of the fly ash concretes. The scaling residue of the fly ash concretes was collected, and in general, the total amount was noticeably less than the usual acceptance limit. The scaling resistance of the concrete was improved by the use of membrane curing instead of the moist curing, especially for the fly ash concretes.

Regardless of the type of the blast-furnace slag used, the slag concretes exhibited considerably more surface scaling than the reference concrete. The surfaces of the slag concretes were given a final visual evaluation of moderate scaling while the total amounts of the scaling residue collected were over the acceptance limit.

In general, for all concretes the amount of surface deterioration increased with the increase of the water-to-cementitious materials ratio, and with the increase of the proportion of the supplementary cementing materials in the mixture.

All high-volume fly ash concretes performed poorly when subjected to the deicing salt scaling test. The superplasticized reference concretes having the same low water-to-cementitous materials ratio as the high-volume fly ash concretes showed some signs of surface deterioration and noticeable amounts of scaling residue.   The poor performance of the high-volume fly ash concretes, and the moderate performance of the reference concretes investigated were probably due, at least in part, to the unsatisfactory quality of the air-void parameters of the concretes, possibly affected by the use of high dosages of the superplasticizer and the finishing process of the slab specimens.

In general, the scaling resistance of concrete appears to be strongly related to the quality of the concrete surface.   The main factor might be the quality of the air-void system at the surface of the concrete specimens.   Other factors like the strength of the concrete and the presence of microcracking at concrete surface may also play a role in the performance of concrete in the deicing salt scaling test.   More research is needed to explain why the concrete incorporating supplementary cementing materials seems to be more affected than the reference concretes.

7   References

1. Carette, G.G. and Malhotra, V.M. (1987) Characterization of Canadian fly ashes and their relative performance in concrete.   **Canadian Journal of Civil Engineering,** Vol. 14, No. 5, 667-682.

2. Malhotra, V.M. (1989) Mechanical properties and freezing and thawing durability of concrete incorporating a ground granulated blast-furnace slag. **Canadian Journal of Civil Engineering,** Vol. 16, No. 2, 140-156.

3. Fernandez, L. and Malhotra, V.M. (1990) Mechanical properties, abrasion resistance, and chloride permeability of concrete incorporating granulated blast-furnace slag. **Cement, Concrete and Aggregates,** Vol. 12, No. 2, 87-100.

4. Carette, G.G. and Malhotra, V.M. (1983) Mechanical properties, durability and drying shrinkage of portland cement concrete incorporating silica fume. **Cement, Concrete and Aggregates,** Vol. 5, No. 1, 3-13.

5. Malhotra, V.M., Painter, K.A. and Bilodeau, A. (1987) Mechanical properties and freezing and thawing resistance of high-strength concrete incorporating silica fume. **Cement, Concrete and Aggregates,** Vol. 9, No. 2, 65-79.

6. Sivasundaram, V., Carette, G.G. and Malhotra, V.M. (1987) Superplasticized high-volume fly ash system to reduce temperature rise in mass concrete. **Proceedings,** Eight International Coal Ash Utilization Symposium, Washington, paper No. 37.

7. Malhotra, V.M. and Painter, K.E. (1989) Early-age strength properties and freezing and thawing resistance of concrete incorporating high-volumes of ASTM Class F fly ash.   **The International Journal of Cement Composites and Lightweight Concrete,** Vol. 11, No. 1, 37-46.

8. Langley, W.S., Carette, G.G. and Malhotra, V.M. (1989) Structural Concrete incorporating high-volumes of ASTM Class F fly ash. **ACI Materials Journal,** Vol. 86, No. 5, 507-514.

9. Sivasundaram, V. Carette, G.G. and Malhotra, V.M. (1989) Properties of concrete incorporating low quantity of cement and high-volumes of low calcium fly ash. **ACI Special Publication SP-114,** Vol. 1, 45-71, (Editor, V.M. Malhotra).

10. Gebler, S.H. and Klieger, P. (1986) Effect of fly ash on the durability of air-entrained concrete. **ACI Special Publication SP-91,** Vol. 1, 483-519, (Editor, V.M. Malhotra).

11. Chojnacki, B. (1975) Partial replacement of portland cement with pelletized slag. **Report IR56,** Research and Development Division, Ontario Ministry of Transportation and communication.

12. Pigeon, M., Perraton, D. and Pleau, R. (1987) Scaling tests of silica fume concrete and the critical spacing factor concept. **ACI Special Publication SP-100,** Vol. 2, 1155-1182, (Editor, J.M. Scanlon).

13. Bilodeau, A. and Carette, G.G. (1989) Resistance of condensed silica fume concrete to the combined action of freezing and thawing cycling and de-icing salts. **ACI Special Publication SP-114,** Vol. 2, 945-970, (Editor, V.M. Malhotra).

14. Bilodeau, A., Carette, G.G. and Malhotra, V.M. (1987) Resistance of concrete incorporating granulated blast-furnace slags to the action of de-icing salts. **Division Report MSL 87-129,** CANMET, Energy, Mines and Resources Canada.

15. Bilodeau, A., Carette, G.G., Malhotra, V.M. and Langley, W.S. (1991) Influence of curing and drying on salt scaling resistance of fly ash concrete. **ACI Special Publication SP-126,** Vol. 1, 201-228, (Editor, V.M. Malhotra).

16. Bilodeau, A. and Malhotra, V.M. (1992) Concrete incorporating high volumes of ASTM Class F fly ashes: Mechanical properties and resistance to deicing salt scaling and to chloride-ion penetration. **ACI Special Publication SP-132,** Vol. 1, 319-349, (Editor, V.M. Malhotra).

17. ACI Manual of Concrete Practice, Part I, American Concrete Institute, Detroit, 1989.

18. Marchand, J., Pigeon, M., Boisvert, J., Isabelle, H.L. and Houdusse, O. (1992) Deicer salt scaling resistance of roller-compacted pavements containing fly ash and silica fume. **ACI Special Publication SP-132,** Vol.1, 151-178, (Editor, V.M. Malhotra).

19. Perenchio, W.F., Whiting, D.A. and Kantro, D.L. (1979) Water reduction, slump loss, and entrained air-void systems as influenced by superplasticizers. **ACI Special Publication SP-62,** 137-155, (Editor, V.M. Malhotra).

20. Plante, P., Pigeon, M. and Saucier, F. (1989) Air-void stability, part II: Influence of superplasticizer and cement. **ACI Materials Journal,** Vol. 86, No. 6, 581-589.

9

Influence of water quality on the frost resistance of concrete

J. STARK and H.-M. LUDWIG
F.A. Finger-Institut für Baustoffkunde, Weinmar, Germany

Abstract
This paper shows that the water quality has an considerable influence on the frost resistance of concrete. Furthermore, it was investigated by using the CF-procedure, that various water qualities affects the scaling behaviour of portland cement concrete and slag cement concrete.
Keywords: blast furnace cement; frost resistance; portland cement; scaling behaviour; water hardness; water quality.

1 Introduction

Until now the quality of the water which is used as test solution in the determination of the frost resistance of concrete has not been defined. In the well-known testing procedures either the use of drinking water is specified or only "water" is mentioned which may as well include the use of distilled water /1/,/2/,/3/,/4/,/5/,/6/,/7/.

Obviously, the influence of the water quality on the results of frost resistance tests was considered to be negligible, although there may be great differences in the quality of water from one location to another. It is not known whether any research has been done in this field.

The results of the second Round Robin Test showed that particularly in the determination of frost resistance the results varied considerably. For instance, in the tests on concrete No. 3 by the CF procedure the results of the amount of scaling after 56 cycles of freezing and thawing varied from 590 g/m^2 to 2190 g/m^2 between the testing institutions.

It was assumed that, in addition to other factors, the different quality of water in the testing institutions might have been the cause of these variations.

Freeze-Thaw Durability of Concrete. Edited by J. Marchand, M. Pigeon and M. Setzer.
Published in 1997 by E & FN Spon, 2–6 Boundary Row, London SE1 8HN, UK.
ISBN 0 419 20000 2.

Therfore, at the Weimar University of Architecture and Building, a Portland cement concrete an a concrete containing blast furnace cement were subjected to the CF procedure using four different qualities of water in order to examine the influence of water on the results of the frost resistance tests.

2 Characterisation of the concretes

As it is possible that the effect of different qualities of water on the test results may also be influenced by the type of cement used, a blast furnace cement was included in the investigations in addition to a portland cement.

The concretes were produced without air-entraining agents and with a high water/cement ratio, which leads to a low frost resistance as previous experience has shown. It is well known that when larger amounts of scaling occur in the CF procedure the variations in a test series will be smaller. Therefore, a higher accuracy may be expected with concretes of poorer quality.

The composition of the test concretes was as follows:

aggregate: natural sand and gravel
                        maximum grain size: 16 mm
cement: PZ 35 F or HOZ 35 L
cement content: 300 kg/m^3
water/cement ratio: 0,65
no air-entraining agent

The results of the tests on fresh and hardened concretes are given in table 1.

Table 1. Results of the tests on fresh and hardened concretes

	Fresh concrete	
	Portland cement concrete	Blast furnace cement concrete
spread table test in cm	47	45
bulk density in kg/dm^3	2,30	2,28
air content in Vol.-%	1,5	1,7

	Hardened concrete	
	Portland cement concrete	Blast furnace cement concrete
bulk density in kg/dm^3	2,28	2,26
compressive strength in N/mm^2 (28 d)	31,9	29,0

In accordance with the specification of the CDF/CF procedure the concretes were stored prior to freezing for one day in the forms, for 6 days under water, and for 21 days in the conditioning room at 65 % relative humidity and 20 °C.

3 Characterisation of testing water

Three different water qualities from the drinking supply sources in the region of Weimar were used. Additionally, distilled water was also used as test solution.

The results of the water analysis are shown in table 2.

It may be seen that the selected water qualities cover the whole range of hardness from very soft to very hard.

Table 2. Results of water analysis

	Destilled water	Water from Nohra	Water from Weimar/centre	Water from Umpferstedt
chloride in mg/l	0	5,2	21,4	45,5
nitrate in mg/l	0	6,6	25,2	46,6
sulphate in mg/l	0,3	25,8	88,7	233,4
silicic acid mg/l	0,1	9,6	12,2	9,0
potassium mg/l	0,01	3,7	3,4	2,7
sodium mg/l	0,03	3,4	9,5	16,0
calcium mg/l	0,07	11,4	86,6	157,9
magnesium mg/l	0	1,8	22,9	27,0
conductivity in µS/cm	2,2	111,1	602,0	963,0
pH-Value	6,1	6,9	7,2	7,6
carbonate hardness in °dH	0	0,5	10,8	11,6
total hardness in °dH	0	1,9	16,4	27,0

4 Results of the tests for frost resistance

The frost resistance was determined on testing five specimens of each of the two types of concrete. As the results of the freeze tests in table 3 show, considerable differences in the extent of damage to the concrete may be observed in dependence on the water used.

Table 3. Amount of scaling after 28 cycles of freezing and thawing

	Amount of scaling after 28 ftc in g/m^2	
	Portland cement concrete	Blast furnace cement concrete
distilled water	202	407
water from Nohra	291	638
water from Weimar	332	817
water from Umpferstedt	489	1502

With portland cement concrete the amount of scaling after 28 cycles of freezing and thawing varies from 202 g/m^2 to 490 g/m^2. The effect of the water quality is even greater with concrete containing blast furnace cement. Here the amount of scaling varies from 407 g/m^2 to 1500 g/m^2. In both cases the amount of scaling increased with the increasing degree of hardness of the water. If we assume a maximum variation coefficient of 20 %, the ratio of mean values for a significant deviation of the series should be 1,5 with a probability of 95 %. It may be stated that the differences between distilled water, soft to medium-hard water and hard water in their effects on portland cement concrete as well as on concrete containing blast furnace cement are statistically significant.

The differences between soft and medium-hard water are within the range of variation in tests.

Figures 1-4 illustrate the extent of scaling for the various testing conditions.

Fig. 1. Scaling detoriation - portland cement concrete/distilled water

Fig. 2. Scaling detoriation - portland cement concrete/water from Umpferstedt

Fig. 3. Scaling detoriation - blast furnace cement concrete/distilled water

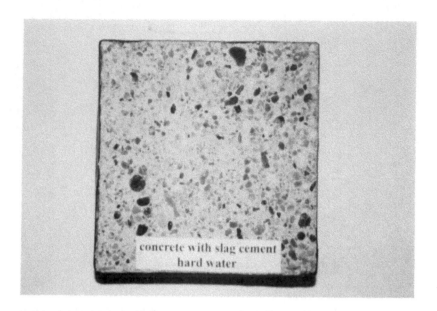

Fig. 4. Scaling detoriation - blast burnace cement concrete/water from Umpferstedt

In general, scaling proceeds in the same way, regardless of the water used (fig. 5 and 6).

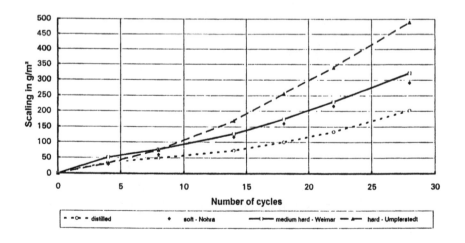

Fig. 5. Scaling as function of cycles - portland cement concrete

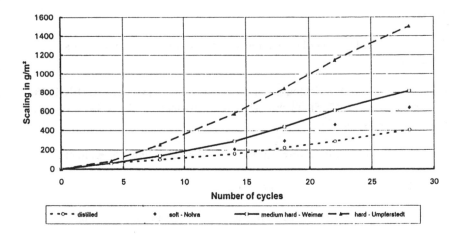

Fig. 6. Scaling as function of cycles - blast furnace cement concrete

A progressive detoriation can be observed on portland cement concrete as well as on concrete containing blast furnace cement, although no release oil was used. The strong initial scaling which is typical of concretes containing blast furnace cement when tested in 3 % NaCl solution did not occur.

5 Discussion

The results presented show that the quality of water may have a considerable influence on the results of frost resistance tests. Before the development of a European standard for the testing procedure this problem should be definitely solved. The investigation described here may be regarded as a first step. Research work has to be continued since some questions still remain unanswered.

For instance, the extent to which an increase in hardness of water led to a higher amount of scaling was unexpected great. Actually, the damage should have been greater in soft water because of the greater extraction of calcium hydrate. A scientific explanation of these experimental findings cannot be given yet. Furthermore, the influence of the water quality in salt solution used for the determination of the freeze-deicing salt resistance has also to be examined. Further investigations should be carried out at several institutions at the same time so as to achieve reliable results.

6 References

1. Swedish Standard SS 137244. Concrete testing - Hardened concrete - Frost resistance.

2. Prüfung von Beton. Empfehlungen und Hinweise als Ergänzungen zu DIN 1048. Deutscher Ausschuß für Stahlbeton. Heft 422, Abschnitt 2.3, S. 12/14, Beuth-Verlag Berlin 1991

3. Setzer, M. J.: Entwurf Prüfvorschrift CDF/CF-Verfahren (unveröffentlicht).

4. ÖNORM B3303. Betonprüfung - Frostbeständigkeit

5. ASTM C666 - 77. Resistance of Concrete to rapid freezing and thawing.

6. TGL 33433/06. Prüfung des erhärteten Betons - Bestimmung des Frostwiderstandes

7. SIA 162/1. Frostwechselverfahren.

10

Efficiency of sealers on the scaling resistance of concrete in presence of deicing salts

K. HAZRATI, C. ABESQUE and M. PIGEON
Centre de recherche interuniversitaire sur le béton, Université Laval, Québec, Canada

T. SEDRAN
Laboratoire Central des Ponts et Chaussées, Paris, France

Abstract
The purpose of this paper is to report on the results of a research program carried out to determine the influence of commercially available penetrating sealers on the deicer salt scaling resistance of different concretes. Three sealers were selected: a silane, an oligomeric siloxane and a polymeric siloxane. They were applied on the concrete surfaces at a constant dosage of 3 m^2/L. The double of the recommended dosage of the silane was also used. Different types of concrete were made: concretes with poor air-void systems, and air-entrained concretes with good or excellent air-void systems. The mixtures were prepared at water/cementitious materials ratios of 0,45 (with normal Portland cement and with silica fume cement) and 0,35 (with silica fume cement). To simulate application of the sealers to concretes previously exposed to deicer salts, the sealers were applied to concrete specimens previously soaked in 0%, 1%, 5%, and 15% sodium chloride solutions for a period of three days. The sealers were also applied to dry as well as humid surfaces. A study of the influence of the application of sealers heated to 40°C and of the influence of weathering of concrete specimens at 40°C before and after application of the sealers was performed. The depth of penetration of the sealers, the water and chloride penetration through the sealed surfaces, and the vapour transmission were determined. The results of the deicer salt scaling tests performed on the various specimens (according to ASTM C672) clearly indicate that the use of sealers reduces the deterioration of concretes that are not properly air-entrained, although it was found that, after a large number of freezing and thawing cycles, this deterioration could be similar to that of untreated surfaces. The results further show that the use of sealers can increase significantly the deterioration of properly air-entrained concretes due to salt scaling. Pre-contamination with salt solutions, even at low concentrations, was found to be very harmful to the performance of the sealers. The sealers, however, can be efficiently applied on humid or microcracked surfaces. The silane was generally found to have the best performance.
Keywords: Durability, concretes, deicer salts, scaling, sealers, silanes, siloxanes, conditions of application.

1 Introduction

The two most important causes of the deterioration of concrete structures in Northern countries are corrosion of the reinforcement and scaling due to freezing in the presence of deicer salts. The frequent use of deicing salts during the winter season (and the

Freeze-Thaw Durability of Concrete. Edited by J. Marchand, M. Pigeon and M. Setzer.
Published in 1997 by E & FN Spon, 2–6 Boundary Row, London SE1 8HN, UK.
ISBN 0 419 20000 2.

exposure to sea water) causes the penetration of salt-laden water into the covercrete. The resulting lowering of the protective alcalinity around the reinforcing steel, together with an adequate supply of oxygen, will lead to accelerated corrosion. The benefits that penetrating sealers can provide against the infiltration of water and chloride ions are well known. They have been commonly used to decrease the risk of corrosion damage for a number of years. Sealers penetrate the concrete surfaces, and react chemically to form a water repellent layer. However, water in the concrete can escape through the treated surface as vapour, and water vapour pressure build up under the sealed surface is prevented. Since deicer salt scaling is also related to the ingress of water and chloride ions [1], the possibility of reducing this type of damage with the use of sealers also exists. The information available in the technical literature on the subject indicates that sealers can in certain cases protect concrete against scaling, but can also in other cases increase the deterioration [2, 3, 4, 5].

2 Objectives of the study

The research project presented in this paper was performed to better understand the effects of penetrating sealers, particularly products from the family of silanes and siloxanes, on the frost-salt durability of both properly air-entrained concretes and concretes with poor air void systems. To simulate actual field conditions, different conditions of application of the sealers were studied.

This investigation also includes a study on the quality of the sealers selected with regard to their basic performance requirements. A good penetrating sealant must, according to [6]:
• penetrate into the concrete surface;
• form an effective barrier to water and chloride ion infiltration;
• provide a "breathable" film that permits the outward transmission of moisture vapor;
• not alter the visual aspect of concrete.

3 Research program

This project was carried out in two phases. In phase I, two basic types of concrete were selected: a standard 0,45 water/cement ratio mixture with a spacing factor of approximately 500 µm, and a similar properly air-entrained mixture (spacing factor ≈ 200 µm). Three different sealers were used at a constant rate of $3m^2/L$: a silane (also applied at double the normal rate), an oligomeric siloxane, and a polymeric siloxane. At the time of testing, these products were generally considered to be the best on the market [7]. The tests were performed on treated and untreated surfaces. Four field conditions of application of the sealers were simulated in the laboratory: on a salt free dry concrete, on a salt free wet concrete, on a salt contaminated dry concrete, and on a salt contaminated wet concrete. The first two represent application on a new structure after the summer, and after a rainy fall. The last two simulate the same weather situations, but for the surface of a concrete structure already subjected to deicer salt applications for a few winters. The salt contamination was achieved with 1%, 5% and 15% NaCl solutions. Two other conditions were also studied: sealers heated at 40°C at the moment of application, and drying of the concrete surface at 40°C for a short time (18 days) to simulate a hot and sunny period. A total of 16 mixtures were produced. The following measurements were performed on the concrete specimens: compressive strength at 28 days, air content and spacing factor, depth of penetration of the sealers, water absorption, vapor transmission, chloride penetration, and freezing and thawing deterioration in the presence of deicer salts.

In phase II the same sealers were used. The mixtures were prepared with normal Portland cement and silica fume cement at water/cementitious materials ratios of 0,45 (both cements), and 0,35 (silica fume cement only). Three sets of concretes were made: concretes with poor air-void systems (spacing factor ≈ 500 µm), and air-entrained concretes with good (spacing factor ≈ 200 µm) and excellent (spacing factor ≈ 100 µm) air-void systems. Two types of coarse aggregates, granitic and limestone (the same as in the first phase) were used. A total of 17 mixtures were produced. In this phase the effects of weathering on the performance of treated concrete surfaces was particularly studied. The concrete specimens were weathered for 9 weeks using two methods: drying at 40°C, and wetting and drying cycles at 40°C. As in phase I, the effects of salt contamination (15%) and the state of saturation of the substrate before the application of the sealers were also studied. The same tests as in phase I were made. The identification code for the mixtures and specimens of the two phases is explained in Figure 1.

4 Materials and mixtures

The mixture proportions and the properties of the freshly-mixed and of the hardened concretes for both phases of the project are given in Tables 1 to 3 (certain phase II mixtures appear twice in Tables 2 and 3, since one large batch was sufficient in certain cases to prepare the specimens required for two test conditions). A normal portland cement (Canadian type 10), and a blended silica fume cement (containing ≈ 8% silica fume) were used. The coarse aggregates consisted of crushed dense limestone or granitic gneiss. The maximum size of the aggregates was 14 mm. The fine aggregate was a granitic sand, and the ratio of fine aggregate/coarse aggregate was kept constant at 40%/60% for the concretes with a water/binder ratio of 0,45, and at 50%/50% for those with a ratio of 0,35. All mixtures but one (489.G35F-DS0) were prepared with an air-entraining admixture. A water reducer was used in the phase II 0,45 mixtures, and a superplasticizer in the 0,35 mixtures.

The following specimens were made from each mixture:
• 4 to 12 75x225x300-mm slabs (2 slabs/test condition) for the scaling resistance tests;
• 6 to 15 75x75x75-mm cubes (2 to 3 cubes/test condition) for the evaluation of the properties of the three sealers;
• one 100x100x400-mm prism for the determination of the parameters of the air-void system;
• 3 to 6 100x200-mm cylinders for the compressive strength tests.

All specimens were cast using a vibrating table. The slabs and cubes were levelled and finished with a wooden trowel. The specimens were then covered with water-saturated burlap and a polyethylene film. They were demoulded after 24 hours, and then cured in lime-saturated water. The 75x225x300-mm slabs were lightly brushed at the surface before the salt scaling tests.

One product from each of the following three different groups of sealers dissolved in organic solvents (white spirits or alcool) was selected: a silane (S1), a oligomeric siloxane (S2), and a polymeric siloxane (S3).

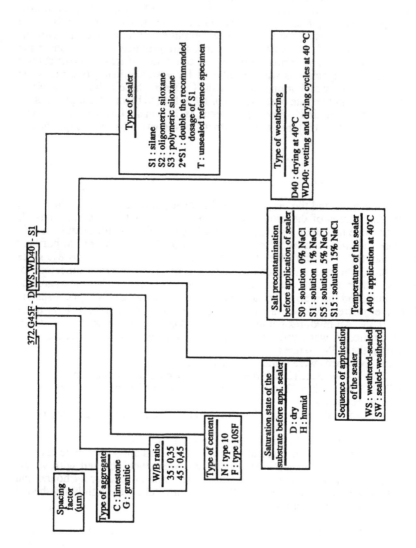

Fig. 1. Identification code for the mixtures and specimens

Table 1. Phase I mixture properties

Mixture	Water (kg/m^3)	Cement (kg/m^3)	Aggregates (kg/m^3) fine	coarse 10mm	14mm	AEA(ml/kg of binder)
509.C45N-DS0	189	421	778	315	736	0.06
237.C45N-DS0	185	412	761	309	720	0.25
701.C45N-DS1	193	430	795	322	752	0.06
131.C45N-DS1	179	398	737	299	697	0.25
853.C45N-DS5	190	423	782	317	740	0.06
116.C45N-DS5	179	397	736	299	697	0.25
584.C45N-DS15	190	423	782	317	740	0.06
126.C45N-DS15	183	407	754	306	713	0.25
445.C45N-HS0	192	426	788	320	746	0.06
194.C45N-HS0	185	412	761	309	720	0.25
431.C45N-HS15	188	417	772	313	730	0.06
220.C45N-HS15	185	410	759	308	718	0.25
694.C45N-DWS.D40	194	430	795	322	753	0.06
128.C45N-DWS.D40	180	400	739	300	699	0.25
738.C45N-DA40	194	430	795	322	753	0.06
135.C45N-DA40	182	406	750	304	710	0.25

Table 2. Phase II mixture properties

Mixture	Water (kg/m^3)	Cement (kg/m^3) type 10	10 SF	Aggregates (kg/m^3) fine	coarse 10 mm	14 mm	AEA*	WR*	SP***
227.C45F-DS0	148	-	328	745	334	779	0.35	4.4	-
554.C45F-DS0	153	-	339	769	345	805	0.05	4.4	-
245.C45F-DS15	147	-	327	742	333	777	0.35	4.4	-
586.C45F-DS15	151	-	336	763	342	798	0.05	4.4	-
213.G45N-DS0	157	348	-	722	326	757	0.12	1.0	-
510.G45N-DS0	163	362	-	731	330	766	0.05	1.0	-
223.G45N-HS0	157	349	-	724	327	760	0.12	1.0	-
216.G35F-DS0	148	-	423	847	340	508	0.5	-	8.5
489.G35F-DS0	152	-	433	887	356	532	-	-	10.0
307.G45N-DWS.D40	159	353	-	733	331	769	0.12	1.0	-
307.G45N-DSW.D40	159	353	-	733	331	769	0.12	1.0	-
247.G45N-DWS.WD40	157	349	-	724	327	760	0.12	1.0	-
247.G45N-DSW.WD40	157	349	-	724	327	760	0.12	1.0	-
127.G45N-DWS.D40	153	339	-	704	318	739	0.31	0.8	-
127.G45N-DSW.D40	153	339	-	704	318	739	0.31	0.8	-
582.G45N-DWS.D40	163	363	-	733	331	769	0.05	1.0	-
582.G45N-DSW.D40	163	363	-	733	331	769	0.05	1.0	-

* ml/kg of binder
** The quantity of water added by the use of SP (58% of the weight) was calculated in the mix design.

Table 3. Concrete properties

Phase I

Mixture	Slump (mm)	Compressive strength (MPa)**	Air content (%)		Spacing factor (µm)*
			Fresh	Hard*	
509.C45N-DS0	85	49,3	2,7	2,0	509
237.C45N-DS0	180	46,1	5,0	4,2	237
701.C45N-DS1	45	60,5	2,0	1,2	701
131.C45N-DS1	162	36,4	6,9	6,6	131
853.C45N-DS5	45	51,8	2,0	0,9	853
116.C45N-DS5	170	39,5	6,7	7,0	116
584.C45N-DS15	50	53,5	2,4	1,3	584
126.C45N-DS15	80	45,6	5,5	4,9	126
445.C45N-HS0	90	54,6	2,7	2,3	445
194.C45N-HS0	120	49,7	5,2	4,0	194
431.C45N-HS15	85	49,3	2,8	1,9	431
220.C45N-HS15	145	45,9	5,0	4,5	220
694.C45N-DWS.D40	105	53,3	1,7	0,8	694
128.C45N-DWS.D40	156	47,1	6,3	6,8	128
738.C45N-DA40	55	58,3	1,5	0,9	738
135.C45N-DA40	110	40,1	5,9	4,5	135

Phase II

Mixture	Slump (mm)	Compressive strength (MPa)**	Air content (%) Fresh	Hard*	Spacing factor (µm)*
227.C45F-DS0	100	42,3	6,2	7,9	227
554.C45F-DS0	100	49,0	3,5	4,5	554
245.C45F-DS15	135	42,4	6,4	7,3	245
586.C45F-DS15	65	50,0	3,2	3,9	586
213.G45N-DS0	95	43,2	6,2	7,0	213
510.G45N-DS0	140	46,8	3,5	2,5	510
223.G45N-HS0	90	42,5	6,0	7,0	223
216.G35F-DS0	120	56,5	8,0	6,0	216
489.G35F-DS0	125	63,6	4,4	3,3	489
307.G45N-DWS.D40	100	48,1	4,5	4,2	307
307.G45N-DSW.D40	100	48,1	4,5	4,2	307
247.G45N-DWS.WD40	110	45,1	5,7	5,4	247
247.G45N-DSW.WD40	110	45,1	5,7	5,4	247
127.G45N-DWS.D40	140	36,6	8,0	8,3	127
127.G45N-DSW.D40	140	36,6	8,0	8,3	127
582.G45N-DWS.D40	115	48,2	3,3	3,0	582
582.G45N-DSW.D40	115	48,2	3,3	3,0	582

** at 28 days (7 in water and 21 days in air), CAN/CSA-A23.2-9C-M90
* ASTM C457

5 Procedures

Figures 2 to 4 describe the methodology used in the laboratory for the preparation of the specimens (including the length of the curing period, the salt precontamination procedure, and the various "weathering" procedures), and for the application of the sealers.

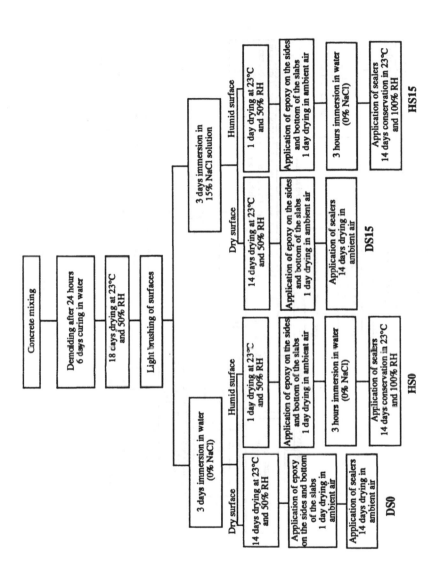

Fig. 2. Preparation and treatment of specimens

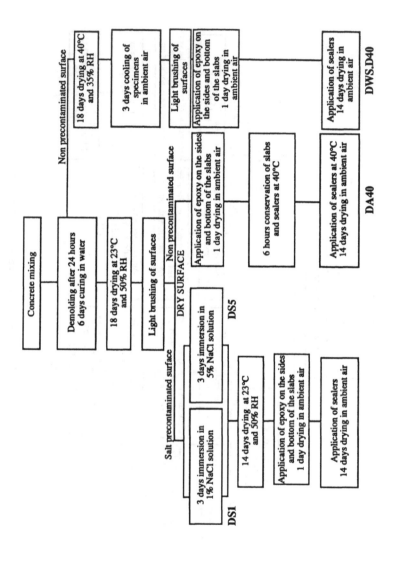

Fig. 3.　Preparation and treatment of specimens

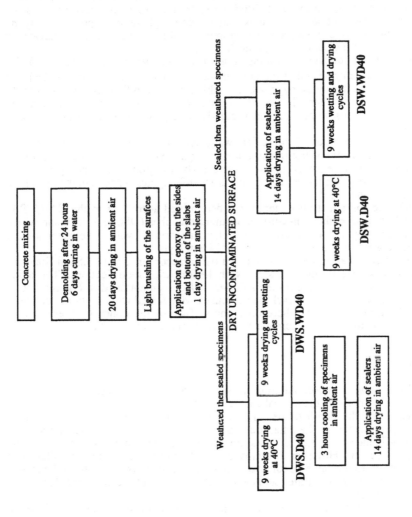

Fig. 4. Preparation and treatment of specimens

6 Description of the tests

The complete tests for the evaluation of the sealer properties (depth of penetration, water absorption, vapor transmission, and chloride penetration) were only performed during phase I because the same sealers were used in the second phase of the project. Only the water absorption and the vapor transmission were reevaluated in phase II (to evaluate the influence of weathering on these properties).

The water absorption and vapor transmission tests were performed on the 75x75x75-mm cubes. After being sealed at the normal rate ($3m^2/L$), the cubes were soaked in a 15% NaCl solution during 28 days, and then dried in a controlled climate room (50% R.H. and 23°C) for another 28 days. This cycle was reversed for the pre-humidified (H) cubes. The cubes were weighed regularly (every 3 days) to determine water movements.

The depth of penetration of the sealers was determined on the same cubes used for the water penetration and vapor transmission tests (at the end of the wetting and drying cycle). The cubes were sawed in two parts and slightly polished. The hydrophobic sealed layer is easily detectable because it has a visible lighter shade than the rest of the cross-section when wetted.

The chloride ion penetration depth was determined according to the CAN/CSA-A23.2-4B-M90 test on the non-salt-precontaminated cubes (DS0 and HS0) and which had already undergone the preceding absorption and drying cycle. The water soluble chloride ion content was measured in the first 25 mm layer.

The deicer salt scaling tests were carried out in accordance with ASTM Standard C672 on the 75x200x300-mm slabs, but with the following modifications. The saline solution was a 2,5% sodium chloride solution. Also, as mentioned in Figures 2 to 4, the perimeter and the bottom of each slab specimen were sealed with an epoxy coating(to simulate the field condition where only one surface of the concrete is exposed to moisture movements). Two specimens were used for each test condition.

7 Results

The results of the visual observations of the surfaces treated with the sealers indicate that these products do not alter the normal appearance of concrete.

The results of the depth of penetration tests are presented in Figures 5 and 6. Considering the scatter of the data, it can be stated that these results indicate no very significant influence of the salt precontamination, of the temperature of application, and of the type of sealer. However, it seems that, on humid surfaces, only the silane can always penetrate to the same depth as on dry surfaces. It is also clear from the results that the sealers tend to penetrate more deeply into the concrete surfaces that have been dried at 40°C (18 days).

The results from the water and chloride ion penetration tests are given in Table 4. They show that sealers are a very effective barrier against water ingress and also against chloride ion ingress (when they are applied on either dry or wet surfaces). However, they tend to loose their efficiency to repel water when they are applied on salt contaminated concrete (particularly the siloxanes), or when the sealed specimens are subjected afterwards to weathering at 40°C by wetting and drying cycles (DSW-WD40). The results of the tests further show that, during the drying period, most of the water absorbed during the wetting period can evaporate through the sealed layer. The three sealers selected for this project were thus generally found to allow efficiently the outward transmisssion of water vapour, except when the tests were performed on specimens subjected to drying at 40°C (D40) or wetting and drying cycles at 40°C (WD40) (which is simply due to the fact that drying during the vapor transmission test was performed at 23°C and not at 40°C). Figures 7 and 8 show two examples of the wetting and drying curves obtained.

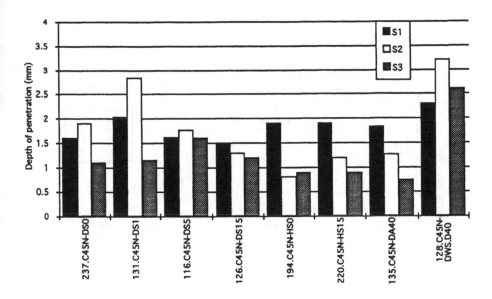

Fig. 5. Depth of penetration of sealers in concretes with good spacing factors

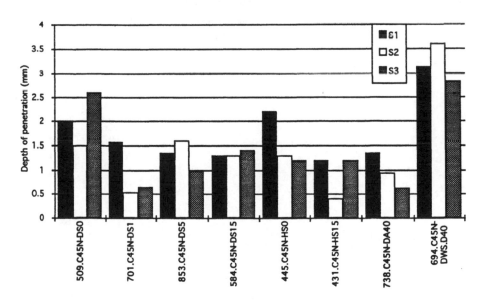

Fig. 6. Depth of penetration of sealers in concretes with insuficient spacing factors

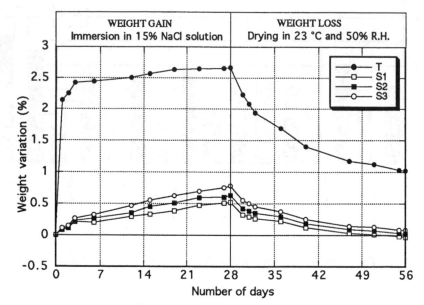

Fig. 7. Wetting and drying test graph
of 237.C45N-DS0 cubes

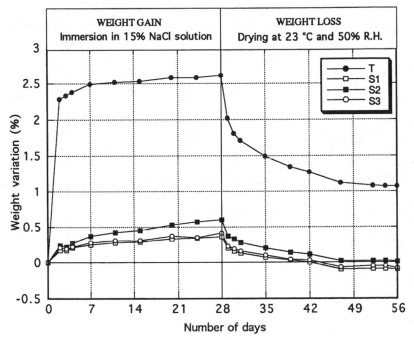

Fig. 8. Wetting and drying test graph
of 509.C45N-DS0 cubes

Table 4. Impermeabilisation properties of sealers

Phase I

Mixture	Water penetration ingress reduction(in %)*			Vapor transmission (in %)*		
Sealer	S1	S2	S3	S1	S2	S3
509.C45N-DS0	86	77	84	127	99	119
237.C45N-DS0	81	76	71	107	99	92
701.C45N-DS1	80	77	76	100	95	99
131.C45N-DS1	77	70	65	97	84	78
853.C45N-DS5	67	65	61	86	85	85
116.C45N-DS5	78	71	67	93	82	74
584.C45N-DS15	55	9	4	98	73	77
126.C45N-DS15	58	14	13	96	74	74
694.C45N-DWS.D40	81	79	66	47	51	54
128.C45N-DWS.D40	75	71	74	62	68	82
738.C45N-DA40	79	79	66	80	86	74
135.C45N-DA40	84	78	65	99	93	81

Phase II

Sealer	S1	S2	S3	S1	S2	S3
227.C45F-DS0	83	81	77	93	98	91
554.C45F-DS0	88	86	83	103	109	104
245.C45F-DS15	62	55	44	126	114	103
586.C45F-DS15	62	57	52	110	108	102
213.G45N-DS0	82	80	82	92	93	98
510.G45N-DS0	85	-	-	94	-	-
216.G35F-DS0	75	71	74	76	78	81
489.G35F-DS0	77	74	76	84	90	95
307.G45N-DWS.D40	80	84	-	28	34	-
307.G45N-DSW.D40	83	82	-	31	25	-
247.G45N-DWS.WD40	88	84	-	97	98	-
247.G45N-DSW.WD40	74	65	-	42	40	-
127.G45N-DWS.D40	83	79	-	39	39	-
127.G45N-DSW.D40	81	78	-	37	33	-
582.G45N-DWS.D40	88	86	-	30	34	-
582.G45N-DSW.D40	86	85	-	33	29	-

Sealer	Chloride ions ingress reduction (in %)*		
	S1	S2	S3
237.C45N-DS0	97	96	94
194.C45N-HS0	97	95	92
509.C45N-DS0	97	96	96
445.C45N-HS0	93	91	77

* Compared to unsealed concrete

Deterioration in the form of "pop-outs" on the concrete surfaces subjected to the deicer salt scaling tests was only observed on the concretes containing the limestone aggregates. This type of deterioration can accelerate the deterioration of the sealed concrete by allowing the water and the chloride ions to infiltrate below the sealed layer.

The results of the deicer salt scaling tests are presented in Table 5. All test specimens were subjected to 50 cycles of freezing and thawing, but, in order to obtain more information, most of them were subjected to an additionnal 50 cycles. Detailed results in the form of graphs showing the mass of residues as a function of the number of cycles are presented in Figures 9 to 41.

Table 5. Mass of scaled-off particles after 50 cycles of freezing and thawing (kg/m^2)

Phase I

Sealer	T	S1	S2	S3	2*S1
509.C45N-DS0	14(16)	0,1(1)	0,5(8)	2,5(8)	0,5(1)
237.C45N-DS0	2,0(2)	0,1(1)	0,6(2)	0,9(2)	0,3(2)
701.C45N-DS1	10,9(15)	1,1(11)	4,1(12)	4,7(11)	-
131.C45N-DS1	0,5(1)	0,4(3)	0,4(2)	0,6(2)	-
853.C45N-DS5	13,0(17)	5,1(13)	5,8(14)	5,5(8)	-
116.C45N-DS5	0,4(1)	1,0(4)	3,3(5)	2,3(3)	-
584.C45N-DS15	7,6	3,2	5,2	5,1	5,9
126.C45N-DS15	0,5	3,4	1,1	1,7	3,9
445.C45N-HS0	5,4(7)	0,0(0)	0,4(5)	2,0(8)	0,1(0,4)
194.C45N-HS0	0,3(0,3)	0,1(0,3)	0,2(1)	0,5(1)	0,1(0,4)
431.C45N-HS15	0,8	1,5	1,5	1,8	1,8
220.C45N-HS15	0,7	1,3	0,1	0,7	1,3
694.C45N-DWS.D40	16,6	0,6	0,9	3,8	-
128.C45N-DWS.D40	0,2(1)	0,4(3)	1,2(4)	1,5(3)	-
738.C45N-DA40	10,7(16)	0,1(11)	0,3(12)	2,7(15)	-
135.C45N-DA40	0,5(1)	0,0(2)	0,2(3)	0,8(2)	-

Phase II

Sealer	T	S1	S2	S3	2*S1
227.C45F-DS0	1,0(1)	0,5(4)	1,0(4)	0,3(1)	
554.C45F-DS0	2,1(3)	0,6(7)	2,8(9)	2,6(6)	
245.C45F-DS15	1,5(2)	3,7(5)	4,2(5)	1,9(2)	
586.C45F-DS15	3,5(7)	3,7(8)	4,3(7)	2,3(5)	
213.G45N-DS0	0,1(0,2)	0,2(2)	0,9(2)	0,5(2)	
510.G45N-DS0	5,3	0,4	-	-	
223.G45N-HS0	0,1(0,1)	0,0(0,3)	0,8(2)	0,5(1)	
216.G35F-DS0	0,1(0,2)	0,0(1)	0,0(1)	0,0(1)	
489.G35F-DS0	0,2(0,2)	0,2(2)	0,2(1)	0,1(1)	
307.G45N-DWS.D40	0,2(0,5)	0,1(2)	0,2(3)	-	
307.G45N-DSW.D40	0,2(0,5)	0,2(2)	0,5(3)	-	
247.G45N-DWS.WD40	0,8(1)	0,0(0,2)	0,0(1)	-	
247.G45N-DSW.WD40	0,8(1)	0,0(2)	0,2(2)	-	
127.G45N-DWS.D40	0,1(0,2)	0,0(2)	0,1(3)	-	
127.G45N-DSW.D40	0,1(0,2)	0,1(2)	0,1(3)	-	
582.G45N-DWS.D40	2,1(6)	0,5(3)	0,3(3)	-	
582.G45N-DSW.D40	2,1(6)	0,4(2)	0,5(2)	-	

in (): mass of scaled-off particles after 100 cycles

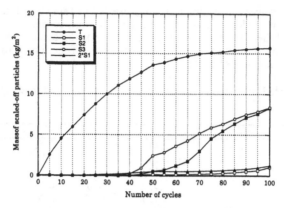

Fig. 9. Mass of scaled-off particules versus number of freezing and thawing cycles for 509.C45N-DS0 specimens

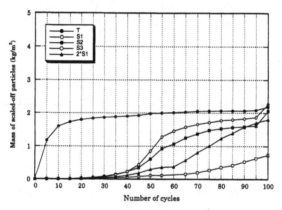

Fig. 10. Mass of scaled-off particules versus number of freezing and thawing cycles for 237.C45N-DS0 specimens

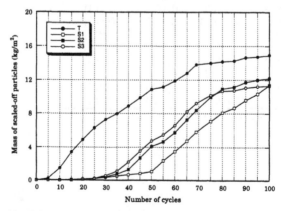

Fig. 11. Mass of scaled-off particules versus number of freezing and thawing cycles for 701.C45N-DS1 specimens

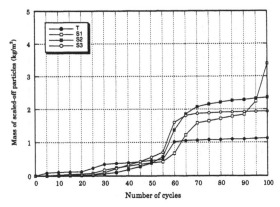

Fig. 12. Mass of scaled-off particules versus number of freezing and thawing cycles for 131.C45N-DS1 specimens

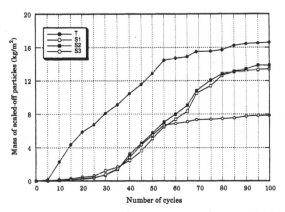

Fig. 13. Mass of scaled-off particules versus number of freezing and thawing cycles for 853.C45N-DS5 specimens

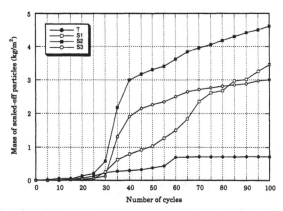

Fig. 14. Mass of scaled-off particules versus number of freezing and thawing cycles for 116.C45N-DS5 specimens

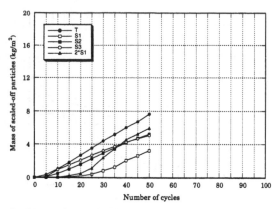

Fig. 15. Mass of scaled-off particules versus number of freezing and thawing cycles for 584.C45N-DS15 specimens

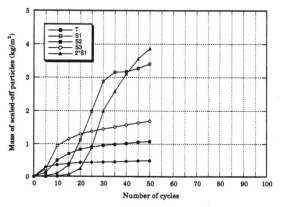

Fig. 16. Mass of scaled-off particules versus number of freezing and thawing cycles for 126.C45N-DS15 specimens

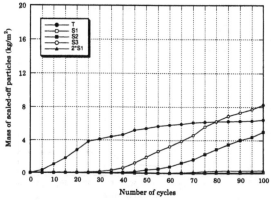

Fig. 17. Mass of scaled-off particules versus number of freezing and thawing cycles for 445.C45N-HS0 specimens

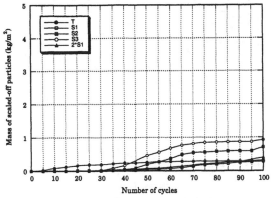

Fig. 18. Mass of scaled-off particules versus number of freezing and thawing cycles for 194.C45N-HS0 specimens

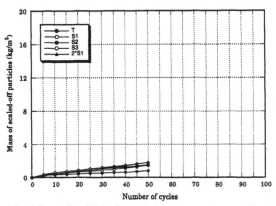

Fig. 19. Mass of scaled-off particules versus number of freezing and thawing cycles for 431.C45N-HS15 specimens

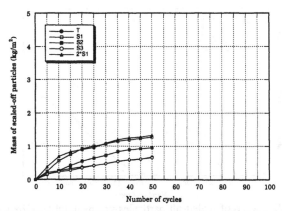

Fig. 20. Mass of scaled-off particules versus number of freezing and thawing cycles for 220.C45N-HS15 specimens

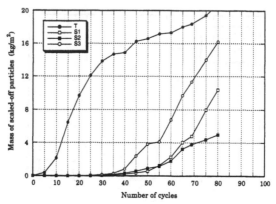

Fig. 21. Mass of scaled-off particules versus number of freezing and thawing cycles for 694.C45N-DSW.D40 specimens

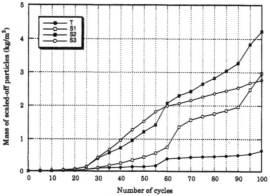

Fig. 22. Mass of scaled-off particules versus number of freezing and thawing cycles for 128.C45N-DWS.D40 specimens

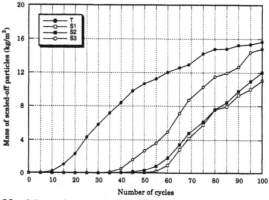

Fig. 23. Mass of scaled-off particules versus number of freezing and thawing cycles for 738.C45N-DA40 specimens

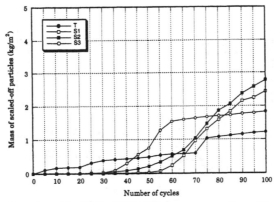

Fig. 24. Mass of scaled-off particules versus number of freezing and thawing cycles for 135.C45N-DA40 specimens

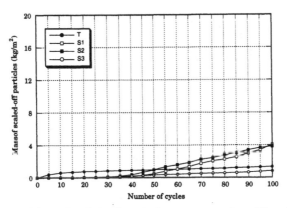

Fig. 25. Mass of scaled-off particules versus number of freezing and thawing cycles for 227.C45F-DS0 specimens

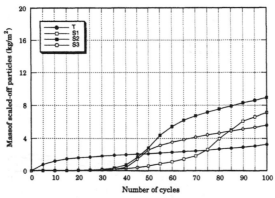

Fig. 26. Mass of scaled-off particules versus number of freezing and thawing cycles for 554.C45F-DS0 specimens

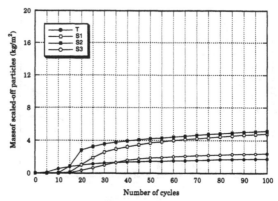

Fig. 27.   Mass of scaled-off particules versus number of freezing and thawing cycles for 245.C45F-DS15 specimens

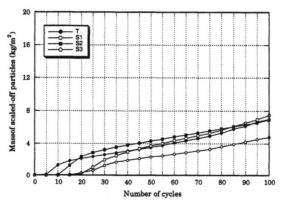

Fig. 28.   Mass of scaled-off particules versus number of freezing and thawing cycles for 586.C45F-DS15 specimens

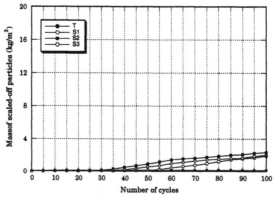

Fig. 29.   Mass of scaled-off particules versus number of freezing and thawing cycles for 213.G45N-DS0 specimens

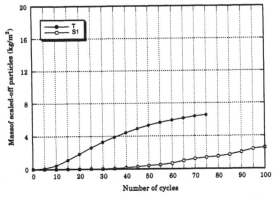

Fig. 30. Mass of scaled-off particules versus number of freezing and thawing cycles for 510.G45N-DS0 specimens

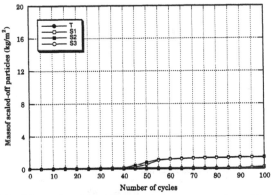

Fig. 31. Mass of scaled-off particules versus number of freezing and thawing cycles for 223.G45N-HS0 specimens

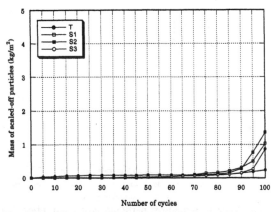

Fig. 32. Mass of scaled-off particules versus number of freezing and thawing cycles for 216.G35F-DS0 specimens

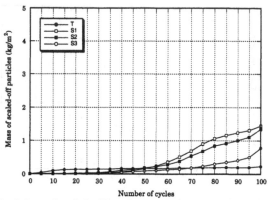

Fig. 33.   Mass of scaled-off particules versus number of freezing and thawing cycles for 489.G35F-DS0 specimens

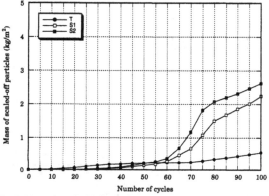

Fig. 34.   Mass of scaled-off particules versus number of freezing and thawing cycles for 307.G45N-DWS.D40 specimens

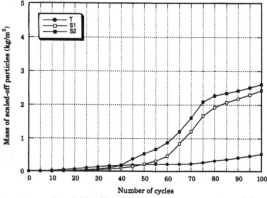

Fig. 35.   Mass of scaled-off particules versus number of freezing and thawing cycles for 307.G45N-DSW.D40 specimens

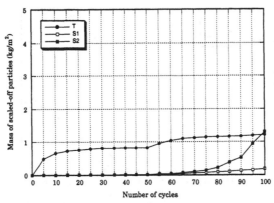

Fig. 36. Mass of scaled-off particules versus number of freezing and thawing cycles for 247.G45N-DWS.WD40 specimens

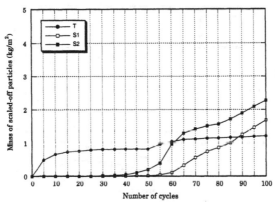

Fig. 37. Mass of scaled-off particules versus number of freezing and thawing cycles for 247.G45N-DSW.WD40 specimens

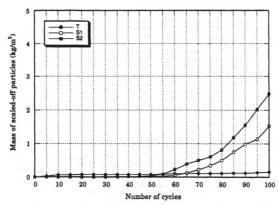

Fig. 38. Mass of scaled-off particules versus number of freezing and thawing cycles for 127.G45N-DWS.D40 specimens

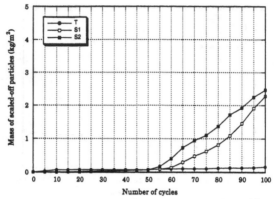

Fig. 39.   Mass of scaled-off particules versus number of freezing and thawing cycles for 127.G45N-DSW.D40 specimens

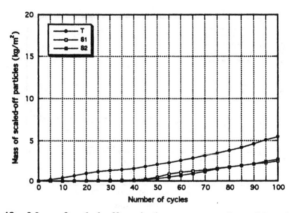

Fig. 40.   Mass of scaled-off particules versus number of freezing and thawing cycles for 582.G45N-DWS.D40 specimens

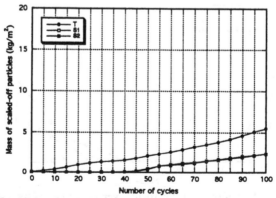

Fig. 41.   Mass of scaled-off particules versus number of freezing and thawing cycles for 582.G45N-DSW.D40 specimens

Generally, concrete is considered resistant to deicer salt scaling if, during the ASTM C672 test, the mass of scaled-off particles (i.e. of the residues collected) is below 1 kg/m^2 after 50 cycles of freezing and thawing. As expected, the mass of scaling residues after 50 cycles for most of the reference (i.e. not treated with sealers) concretes with a poor air void system is high (above 5 kg/m^2). The result for mixture 431.C45N-HS15-T, however, is surprising, since the mass of scaled-off particles after 50 cycles is only 0,8 kg/m^2. This indicates that the use of air-entraining admixtures, even when the dosage is low, can contribute, in certain circumstances which are not completely clear, to enhance significantly the deicer salt scaling resistance. The mass of residues was also found to be very low for the non-air-entrained silica fume mixture with a water/binder ratio of 0,35 (489.G35F-DS0-T). This confirms the good durability of such "high performance" concretes. For the air-entrained reference concretes, the results are generally very good (i.e. the mass of residues is <1 kg/m^2), although the mass of residues for a few mixtures, such as 237.C45N-DS0-T, is relatively high (more than 1 kg/m^2 after 50 cycles). Since this deterioration was found to occur mainly in the first 10 cycles, it was thus probably due to a surface problem related to the finishing operations.

The weathering procedures, such as drying at 40°C, and the salt precontamination, were not found to have any significant influence on the scaling resistance of the various reference concretes. This is uncommnon, particularly as regards drying at 40°C, and remains to be explained.

The results in Table 5 show that the use of sealers can reduce very significantly the scaling deterioration of concretes with a poor air void system. The detailed results (Figures 9 to 41) further indicate that the protection offered by the use of sealers tends to disappear after a certain number of cycles, and that the silane (S1) is generally better than the other two sealers in this respect. The use of this sealer at double the normal rate, however, was found to have no very significant influence (Figures 9, 15, 17 and 19). In fact, the deterioration is always slightly higher with the double than with the simple rate.

The influence of sealers on the scaling resistance of the air-entrained concretes is very different. After a certain number of cycles, the specimens generally suffer a sudden deterioration, and, at the end of the cycles, the mass of scaling residues is often higher for the concretes treated with the sealers than for the untreated ones (see Figures 9 to 41).

As previously mentioned, sealers can be applied on dry surfaces (D), as well as on wet surfaces (H). This does not seem to have any significant influence on the scaling resistance of concrete (see for example Figures 9 and 17).

The salt contamination of the concrete surfaces (even at a low concentration) before the application of the sealers clearly has a negative effect on the performance of the sealers. The precontaminated sealed concretes all exhibit considerably more surface scaling in the first 30 to 50 cycles than the non-contaminated sealed concretes (see for example Figures 9 to 16). The test results clearly show that, when there is salt contamination, the protection offered by sealers to concretes with poor air void systems is reduced significantly, and that the negative influence of sealers on air-entrained concretes generally occurs after a smaller number of cycles, depending on the degree of contamination. The results also show that the deterioration at the end of the cycles for the air-entrained concretes treated with the silane (S1 or 2*S1) is often larger than that of the same concretes treated with the other sealers (figures 10, 16, 18 and 20).

Unlike the mixtures with a water/binder ratio of 0,45, the air-entrained and non-air-entrained sealed concrete mixtures with a water/binder ratio of 0,35 (216.G35F-DS0 and 489.G35F-DS0) have similar behaviors (Figure 32 and 33). Both mixtures perfom like good quality concretes for which the use of sealers affects adversly the deicer salt scaling resistance. However, the mass of residues for the sealed specimens is reduced compared to that of the air-entrained mixtures with a water/binder ratio of 0,45, and the deterioration occurs after a larger number of cycles.

As previously mentioned, drying at 40°C did not have any significant effect on the salt scaling resistance of the reference concretes. The results also show that the influence of sealers on the salt scaling resistance was similar for these artificially "weathered" concretes and for the non-weathered concretes (Figures 21,22, 34 to 41). Furthermore, in such cases, as can be seen in the Figures, the sequence of application of the sealers (i.e. before (DSW-D40) or after (DWS-D40) drying) had no apparent effect. However, the sequence of application of the sealers on the concrete surfaces did seem to influence the results when the specimens were weathered by wetting and drying cycles (see Figures 36 and 37). In this case, the positive effect of the sealers was reduced when the sealed specimens were "weathered".

According to the test results (Table 5, and Figures 23 and 24), heating of the sealers at 40°C before application on the concrete surfaces (DA40) has no apparent influence on their salt scaling performance.

8 Discussion

The results of the characterization tests performed on the cubes show that the silane and the siloxanes prevent very efficiently the ingress of water and chloride ions into concrete, and can thus protect the reinforcing steel from corrosion. They also allow the vapor transmission and do not change the external appearance of the concrete surface. However, the use of sealers on concrete surfaces tends to modify very significantly the salt scaling resistance. Figures 9 to 41 shows that the deterioration of the concretes treated with sealers does not increase linearly with the number of cycles, as it is often the case with untreated concretes [8]. After a certain number of cycles, which can vary with the type of the sealer and the conditions of application, the deterioration increases very suddendly. As confirmed by the visual observations of the test specimens, this deterioration corresponds to the loss of a significant layer of concrete. This suggests that the damage tends to occur under the layer of concrete penetrated by the sealer.

Sealers can have a beneficial effect against scaling by reducing the degree of saturation of the surface layers of concrete. Sealers prevent, at least to some degree, the ingress of water into that part of the concrete penetrated by the sealer. Thus, initially, the surface layers of concrete are probably not in a state of critical saturation. It can also be hypothecized that the internal, so-called "osmotic", pressures created by the presence of the chloride ions [9] are reduced. On the other hand, sealers can also have a negative effect. Water movements are hindered by the presence of an impervious concrete layer. When the unfrozen water (in the small pores) of the concrete under the sealed layer is drawn to the ice front at the surface of the concrete, significant internal pressures can be formed under this impervious layer.

Based on these considerations, the following mechanism is suggested to explain the effect of silanes and siloxanes on the frost-salt durability of concrete. The scaling of sealed concrete can be divided into three stages (Figures 42 and 43):

•Stage 1: gradual saturation under the sealed layer up to critical saturation. This stage extends from the first freezing and thawing cycle up to a limit called "the efficiency threshold" (or "the protection threshold") of the sealer. In this stage the use of the sealer is beneficial and scaling of the concrete is considerably reduced.

•Stage 2: this is the stage where the use of the sealer has a negative effect. The deterioration of the concrete increases very suddenly. Damage occurs, not at the surface, but under the layer of concrete penetrated by the sealer (Figure 43). This layer can be very rapidly separated from the bulk of the specimen. This sudden deterioration is most probably simply due to basic frost effects in an area which has become critically saturated.

•Stage 3: once the sealed concrete layer is completly eliminated (stage 2), the scaling behavior is similar to that of an unsealed concrete.

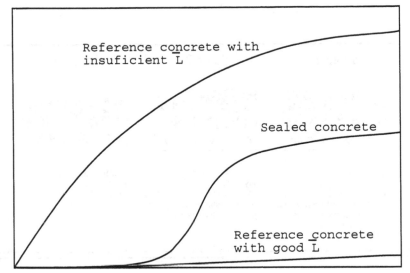

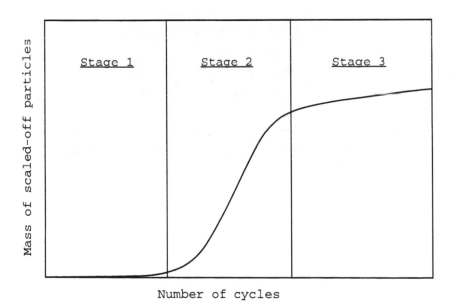

Fig. 42. Typical freeze-thaw durability curves of sealed concretes

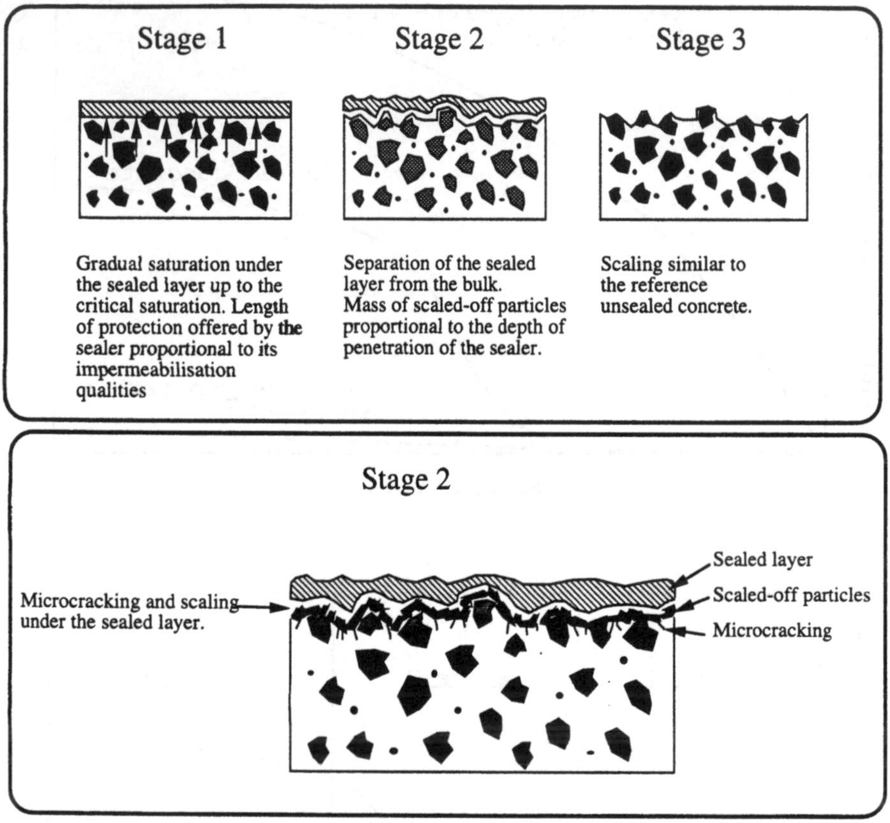

Stage 1	Stage 2	Stage 3
Gradual saturation under the sealed layer up to the critical saturation. **Length** of protection offered by **the** sealer proportional to **its** impermeabilisation qualities	Separation of the sealed layer from the bulk. Mass of scaled-off particles proportional to the depth of penetration of the sealer.	Scaling similar to the reference unsealed concrete.

Stage 2

Microcracking and scaling under the sealed layer.

Sealed layer

Scaled-off particles

Microcracking

Fig. 43. Explanation of the three stages of deterioration of sealed concretes submitted to deicing salt scaling tests

The length of stage 1 depends on the quality of the sealer. That is probably why the silane, considered the best of the three products selected, generally protected the specimens from water saturation and penetration of chloride ions for the longest period. When the salt pre-contamination is high, the length of stage 1 is reduced for all sealers and all types of concrete. From the results of the wetting and drying tests (Table 4), it can be assumed that salt attacks the integrity of the sealed layer. Paths are opened for the water and chloride ions to penetrate into the concrete and contribute to the saturation of the concrete directly under the sealed layer. Any defect in the sealed layer such as cracks or "pop-outs" of the aggregates (after the application of the sealer) should affect the length of stage 1.

Although, because of the scatter of the data, the test results do not demonstrate this very clearly, the intensity of the damage in stage 2 should depend on the depth of penetration of the sealer. Considering that a loss of 1 kg/m^2 corresponds approximately to a concrete thickness of 0,4 mm, this could explain, for instance, the severe deterioration of 126-C45N-DS15 sealed with the silane (S1 ou 2*S1) (Figure 16). Because of their relatively small size molecules, silane sealers normally penetrate deeper into the concrete compared to the siloxanes sealers (S2 et S3) whose molecules usually consist of 2 to 10 silane molecules grouped together. In the case of the mixtures with a water/binder ratio of 0,35 (with a good or a poor air void system), the damage caused during stage 2 was reduced, because the sealers did not penetrate as deep into these concretes.

Drying tends to modify the concrete porosity [10] and, because of humidity gradients, to create microcracks at the surface. The results of the scaling tests on the concrete specimens weathered by drying at 40°C before or after the application of sealers (DWS.D40 or DWS.D40) were quite similar to those of the specimens dried at room temperature (with normally less microcraking). Sealers of the type of silanes and siloxanes thus seem to waterproof efficiently this type of surface. In the case of weathering by wetting and drying cycles at 40°C after the application of the sealers (DSW-WD40), the silane and the siloxanes were found to have a poorer performance. The results of the wetting and drying performed on the 75 mm cubes show that DSW-WD40 specimens absorbed more water than their DWS-WD40 counterparts. It can thus be assumed that wetting and drying cycles at 40°C damages the sealed layer, whereas simple drying obviously does not.

Heating at 40°C of the sealers before their application (compared to the application at 23°C) had no apparent effect on the scaling test results. It is possible that, at 40°C, the solvent evaporation of the product is high and, even though the viscosity of the product is reduced, it does not thus penetrate more than at 23°C (see Figures 5 and 6).

9 Conclusions et recommendations

The test results described in this paper show that sealers such as silanes and siloxanes can have a significant influence on the resistance of concrete surfaces to deicer salt scaling.

Sealers appear to be useful to reduce deicer salt scaling in only one case: on new (non salt-contaminated), and insufficiently air-entrained, concrete. The protection offered by sealers, which is a function of the nature of the sealer, is limited in time, and it is not possible to estimate with a sufficient degree of precision the length of the period of efficiency under field conditions.

When used on concretes which are adequately salt scaling resistant, sealers do not add any supplementary protection. On the contrary, the use of sealers often tends to increase the scaling deterioration. It should be noted, however, that this conclusion is based on tests carried out in the laboratory, where the surface of the specimens is continuously covered with salt water.

Salt pre-contamination of the concrete reduces very significantly the efficiency of sealers. It thus reduces the protection against scaling offered to poorly air-entrained concretes, and accelerates the deterioration of air-entrained concretes. Any contaminated layer should thus be removed before sealer application.

Sealers applied on saturated surface-dry concretes seem to be as efficient as when they are applied on dry concretes, even if they penetrate a little less.

The temperature of the silane and the siloxanes at the time of application does not seem to have a very significant influence.

The silane sealer was generally found to have the best performance, i.e. to reduce scaling of new (non salt-contaminated) and poorly air-entrained concrete.

10 Acknowledgments

The authors are grateful to the Fonds FCAR from the Québec government, and to the Québec Ministry of Transportation, for their financial support for this project. The authors also wish to thank Daniel Vézina from the Ministry of Transportation for his advice and his technical help.

11 References

1.   Verbeck C.J. and Klieger P. (1957), *Studies of salt scaling of concrete*, Highway Res.Board Bulletin, n°150, 1-13.
2.   Vesikari E. (1983), *Prevention of salt action on concrete by use of surface sealants*, Nordic Concrete Research, n°2 Dec. 83, 205-214.
3.   Bradbury A. and Chojnacki B. (1985), *A laboratory evaluation of concrete surface sealants*. MI-79 report, Ministry of Transp.and Com. of Ontario.
4.   Rollings R.M. and Chojnacki B. (1988), *A laboratory evaluation of concrete surface sealants phase 2*, MI-128 report, Ministry of Transp.and Com. of Ontario.
5.   Perenchio W.F. (1988), *Durability of concrete treated with silanes*, Concrete International, Nov. 1988, pp.34-40.
6.   Pfeifer D.W. and Scali M.J. (1981), *Concrete for protection of bridge structures*, National Cooperative Highway Research Program Report 244, TSB, Washington, D.C., 130 p.
7.   Letourneau J.G. and Vézina D. (1988), *Evaluation des imperméabilisants à béton destinés aux ouvrages d'art*, RTQ 87-12 report, Ministère des Transports du Québec.
8.   Pigeon M. (1989), *La durabilité au gel du béton*, Materials and Structures, vol.22, pp.3-14.
9.   Powers, (1975)......
10.   Sellevold E.J. and Farstad T. (1991), *Frost/Salt testing of concrete:Effect of test parameters and concrete moisture history*, Nordic Concrete Research, vol. 10 pp. 121-138.

Hazrati K. (1993), *Efficacité des scellants face au problème d'écaillage des bétons en présence de sels fondants*, Master degree thesis, Departement of Civil Engineering, Laval University, Ste-Foy, Québec, 110 p.

Part III

Assessing the frost durability and the deicer salt scaling resistance of concrete

11

Reliability of the ASTM C 672
test procedure

M.C. LAROCHE, J. MARCHAND and M. PIGEON
Centre de recherche interuniversitaire sur le béton, Université Laval, Québec, Canada

Abstract
Specimens from various sidewalks in the City of Québec (where the winter conditions are very severe) were tested for deicer salt scaling resistance using the ASTM C 672 test procedure. Two series of concretes were selected for the tests: a first series of six concretes showing little scaling after six winters and a second series of four concretes showing severe scaling damage after only three winters of exposure. In addition to these field concretes, three similar mixtures were prepared in the laboratory and tested for scaling resistance to analyze the influence of the drying period before the tests. The results show that deicer salt scaling test procedures such as that described in ASTM C 672 are not too severe and allow to distinguish with a fair degree of reliability between those concretes which are scaling resistant and those which are not. The results further show that drying at 40° C during 14 days before the freezing and thawing cycles is most probably not a condition which is representative of field conditions, and that such a drying procedure should not be used before performing deicer salt scaling tests.
Keyword: Deicers, scaling, frost durability, test procedures, field behavior.

1 Introduction

The regular use of deicer salts during snow-removal operations is one of the main causes of the rapid deterioration of many concrete structures. Salts increase the risk of reinforcing steel corrosion as well as the damage due to freezing and thawing cycles.

Since the early fifties, the use of deicer salts has largely contributed to increase the number of structures damaged by surface scaling. This has prompted many investigators to study the basic mechanisms of scaling due to freezing in the presence of deicer salts, and to try to find ways to increase the resistance of concrete to this type of damage. In many cases, these investigations were performed in the laboratory using standardized procedures.

The reliability of these standardized laboratory test procedures is now being questioned. The few investigators that have systematically compared the results of laboratory tests to field performance have concluded that laboratory tests are generally too severe and do not allow to determine if a given concrete will be durable or not under field exposure conditions [1-3].

Freeze-Thaw Durability of Concrete. Edited by J. Marchand, M. Pigeon and M. Setzer.
Published in 1997 by E & FN Spon, 2–6 Boundary Row, London SE1 8HN, UK.
ISBN 0 419 20000 2.

The present study was performed to examine the reliability of the most commonly used test procedure to assess the resistance of concrete to deicer salt scaling, the ASTM C 672 test method, and to allow a better evaluation of the ability of this test to predict concrete durability. The resistance of various concretes to scaling was measured in the laboratory and compared to field performance.

2 Research program

In order to be able to compare the results of laboratory tests to field performance, specimens were taken from various sidewalks in the City of Québec where the winter conditions are very severe (low temperatures, frequent freezing and thawing cycles and large use of deicer salts). Two series of concretes were selected:

- a first series of six concretes showing little scaling after six winters;
- a second series of four concretes showing severe scaling damage after only three winters of exposure.

Information concerning the first series with regard to mixture proportions, fresh concrete properties, placing, finishing and curing techniques was available, since these concretes were cast as part of a research program on the stability of the air void system [4]. For the second series, however, only general and approximate data is available. In addition to the field concretes, three similar mixtures were prepared in the laboratory and tested for scaling resistance to analyze the influence of the air void spacing factor and of the drying period before the tests.

3 Experimental

3.1 Materials
The first six concretes were produced in three different plants (simply referred to as A, B and C in this paper), each producer using a different ordinary Portland cement (Canadian type 10). Table 1 gives some of the basic chemical information for these three cements. Unfortunately, it was impossible to obtain the chemical analysis of the cements used for the second series of four concretes. The information concerning the cement used for the preparation of the laboratory concretes is presented in Table 2. All concretes (field and laboratory) were made with similar fine and coarse aggregates (natural granitic sand and crushed limestone from the Québec area). A lignosulfonate water reducer was used for the first series of field concretes and for the laboratory concretes, and a hydroxylated polymer for the second series of field concretes. A synthetic detergent air-entraining admixture was used for all mixtures but A 10-02, A 10-04 and C 10-06 (see Table 3 in the following sub-section for the mixture description) in which salts of sulfonated hydrocarbon were utilized. Mixtures B 10-05 and C 10-06 also contained different amounts of a naphtalene based superplasticizer.

3.2 Mixture composition
The composition of all concretes, field and laboratory, is given in Table 3. The nominal water/cement ratio of all field concretes is equal to 0.45. This is a requirement of the Canadian Code of practice for concretes exposed to severe winter conditions and deicer salt applications. In most cases, however, these concretes were retempered with water to increase the slump before placement and, sometimes, an additional dosage of the air-entraining admixture was added. The water/cement ratio was thus often higher than 0.45.

Table 1. Available information on the three cements of series 1

| Characteristic | Type 10 cement | | |
	A	B	C
Blaine (m^2/kg)	370	324	416
C$_3$A (%)	8.9	7.5	5.3
Total alkali (%)	1.0	0.7	1.0
L.O.I. (%)	0.9	2.7	2.3
Limestone filler	no	yes	yes

Table 2. Chemical and mineralogical compositions of the cement used in series 2

Chemical composition

Constituent	Proportion (%)	Constituent	Proportion (%)
SiO$_2$	20.6	MgO	2.7
Al$_2$O$_3$	4.9	Na$_2$O	0.22
Fe$_2$O$_3$	2.4	K$_2$O	0.48
CaO	62.8	L-O-I	1.2
Free lime	0.4	Res. imp.	—
SO$_3$	2.7		

Mineralogical composition (calculated according to Bogue's equations)

Constituent	Proportion (%)	Constituent	Proportion (%)
C$_3$S	56.6	C$_3$A	8.9
C$_2$S	16.4	C$_4$AF	7.3

For the first series of field concretes, the following summarizes what is known concerning retempering:

- water was added to mixtures A 10-01 and A 10-02 before the last series of tests on the fresh concrete which were carried out at 85 minutes (after the initial water-cement contact);
- water and air-entraining admixture was added at 55 minutes to mixture A 10-03;
- a large amount of water was added to mixture A 10-04 at 60 minutes, and air-entraining admixture was added at 80 minutes;
- water was added to mixture B 10-05 (superplasticized) after the last series of tests (at 85 minutes) to increase the slump to 80 - 100 mm;
- air-entraining admixture was added at 65 minutes to mixture C 10-06 to increase the air content to the required level.

Table 3. Mixture characteristics

Mixture	Cement (kg/m³)	Water (kg/m³)	Sand (kg/m³)	Stone (kg/m³)	AEA (mL/m³)	WRA (mL/m³)	SP (mL/m³)
Durable field concretes — Series 1							
A10-01	368	165	798	1000	110	1619	—
A10-02	368	165	798	1000	129	1619	—
A10-03	368	165	798	1000	44	1619	—
A10-04	368	165	798	1000	44	1619	—
B10-05	375	165	895	933	109	1650	1613
C10-06	350	157	817	990	130	1540	1330
Non-durable field concretes — Series 2*							
N10-01	365	165	800	960	91	876	—
N10-02	365	165	800	960	91	876	—
N10-03	365	165	800	960	91	876	—
N10-04	365	165	800	960	91	876	—
Laboratory concretes — Series 3							
D10-01	363	172	730	1080	44	363	—
D10-02	363	160	742	1080	18	363	—
D10-03	363	157	748	1080	7	363	—

AEA: Air-entraining agent; WRA: Water-reducing agent; SP: Superplasticizer
* According to plant records

The water/cement ratio of the three laboratory concretes is also equal to 0.45. The dosage of the air-entraining admixture was varied to obtain air void spacing factors of approximately 200 μm, 300 μm and 400 μm respectively.

3.3 Mixing, placing and finishing techniques
Two mixing procedures were used for the field concretes: mixing in the ready-mix truck (producers A and B), and in-plant mixing in a premix unit (producer C and second series of concretes).

A pan-type mixer was used for the laboratory concretes. For these mixtures, the fine aggregate was first homogenized in the mixer for about one minute. The cement was added to the fine aggregate and mixed until it was uniformly distributed, and then the mixing water together with the admixtures (separately, each with a little of the mixing water) were added. After two minutes of mixing, the coarse aggregate was finally added and mixing continued for three more minutes.

The laboratory test specimens were cast in two layers, each one being vibrated for 20 seconds. The top surface was lightly finished with a wooden trowel.

3.4 Curing
A membrane forming curing compound was used for all field concretes. The laboratory test specimens were covered with wet burlap for the first 24 hours after casting and then placed in lime saturated water for the required period of time (i.e. 7, 28 and 90

days for the compressive strength measurement specimens and 14 days for the scaling test specimens).

3.5 Conditioning

The test specimens taken from the various sidewalks were all simply stored in the laboratory until testing. For the laboratory concretes, three drying procedures were used (immediately following curing): normal drying at 50% R.H. and 23° C for 14 days as specified in ASTM C 672, oven drying at 40° C for 14 days (such harsh drying is considered possible under field conditions during the summer) and drying for 90 days at 50% R.H. and 23° C.

3.6 Test procedures

The air void characteristics of all field and laboratory concretes was determined using the modified point count method of ASTM C 457. The compressive strength of the field concretes was measured on dry cores (with length/diameter ratios ranging between 1.09 and 1.63), according to ASTM C 39. As required, the results were corrected to take the length/diameter ratio into account. The compressive strength of the laboratory concretes was determined on 100 mm in diameter and 200 mm in length standard cylinders.

The procedure described in ASTM C 672 was used for all deicer salt scaling tests. For each field concrete, three 75 mm long and 195 mm in diameter cores were tested. These cores were taken from the undamaged part of the surface of the sidewalks. For the laboratory concretes, two 75 x 225 x 300 mm slabs were used for each test condition. At the end of the drying period following curing, the test surface of all specimens was covered with pure water for three days, and then immediately replaced with a 3% sodium chloride solution for the freezing and thawing cycles. This solution was renewed every 5 cycles when the residues were collected and weighed and the surface rated according to the standard.

4 Test results

4.1 Fresh concrete properties

The slump and the air content measured on the fresh concrete for all field concretes are given in Table 4. All slump values for the first series of six "durable" concretes were determined 85 minutes after the initial water-cement contact. These values range between 45 mm and 70 mm, except for mixture A 10-04 (135 mm) which was retempered with a large amount of water at 60 minutes and with air-entraining admixture at 80 minutes. The slump values and air contents for the second series of four "non-durable" concretes are not known and only the target values are shown. Table 4 also gives the air content and the slump of the laboratory concretes, measured 10 minutes after the initial water-cement contact. The air content varies between 6.2% and 3.1% since the dosage of the air-entraining admixture was adjusted to produce concretes with spacing factors varying from 200 μm to 400 μm. The slump however, is approximately constant at 125 to 150 mm for these three laboratory concretes.

4.2 Compressive strength

The results of the compressive strength tests are presented in Table 5 for the field concretes and Table 6 for the laboratory concretes. The values range from 38 to 57 MPa which is quite typical of such field concretes with variable air contents and that are often retempered. At 28 days, the strength of the laboratory concretes varies normally between 38 MPa (for the mixture with the largest volume of air) and 48 MPa (for that with the lowest volume).

Table 4. Fresh concrete properties

Mixture	Air content (%)	Slump (mm)
Durable field concretes — Series 1		
A10-01	6.8	60
A10-02	5.1	50
A10-03	5.8	45
A10-04	4.6	135
B10-05	4.5	45
C10-06	4.8	70
Non-durable field concretes — Series 2*		
N10-01	6 to 8	80 ± 20
N10-02	6 to 8	80 ± 20
N10-03	6 to 8	80 ± 20
N10-04	6 to 8	80 ± 20
Laboratory concretes — Series 3		
D10-01	6.2	135
D10-02	4.6	150
D10-03	3.1	125

4.3 Air void characteristics

The characteristics of the air void system of the field concretes are presented in Table 5 and those of the laboratory concretes in Table 6. The spacing factor of the field concretes is generally satisfactory, the values ranging between 135 μm and 282 μm. For concretes exposed to severe winter conditions, the Canadian Code of Practice states that the average value of the air void spacing factor shall not exceed 230 μm and no individual value shall exceed 260 μm. The spacing factors of the laboratory concretes are 204 μm, 299 μm and 412 μm respectively, i.e. close to the target values.

4.4 Resistance to deicer salt scaling

The deicer salt scaling test results are summarized in Table 7 (field concretes) and Table 8 (laboratory concretes) which give both the amount of residues and the visual rating after 50 cycles. Fig. 1 and 2 also show, for the first series of durable field concretes and the second series of non-durable field concretes respectively, the amount of residues as a function of the number of cycles.

With the exception of concrete A 10-04, the results in Table 7 indicate that the amount of residues after 50 cycles is extremely small for all durable field concretes (0.04 to 0.11 kg/m^2). Despite its very good field performance, the value for concrete A 10-04 (which was retempered with a large amount of water) is approximately 10 times the average value for the other concretes of the series. Its value is, however, still lower than the 1 kg/m^2 limit suggested in the Swedish Standard SS 13 72 44.

Table 5. Compressive strength results and air-void characteristics of field concretes

| Mixture | Air-void characteristics | | | Compressive strength |
	Air content (%)	Spec. surface (mm^{-1})	Spacing factor (μm)	(MPa)
Durable field concretes — Series 1				
A10-01	8.6	22.5	135	38
A10-02	5.2	18.9	252	45
A10-03	6.7	23.0	186	54
A10-04	6.3	21.1	214	48
B10-05	4.6	22.2	229	50
C10-06	7.5	21.8	178	50
Non-durable field concretes — Series 2				
N10-01	5.2	21.9	213	41
N10-02	7.1	14.5	263	46
N10-03	5.2	15.7	282	57
N10-04	5.7	24.1	192	42

Table 6. Compressive strengths and air-void characteristics of laboratory concretes

| Mixture | Air-void characteristics | | | Compressive strength | | |
	Air content (%)	Specific surface (mm^{-1})	Spacing factor (μm)	7 d. (MPa)	28 d. (MPa)	90 d. (MPa)
D10-01	5.8	21.1	204	31	38	45
D10-02	5.4	16.3	299	34	42	46
D10-03	3.8	13.4	412	37	48	50

For the non-durable field concretes, the results in Table 7 show that, on the average, the amount of residues after 50 cycles of freezing and thawing in the laboratory is much higher than the amount for the durable concretes. However, all values are again lower than the 1 kg/m^2 limit, with the exception of N 10-04 for which the amount of residues is only slightly higher than the 1 kg/m^2 limit. Fig. 1 shows that the scaling process for the two most deteriorated concretes of this series is not linear, about half of the deterioration occurring during the first 10 cycles, after which the rate of deterioration becomes similar to that of the other concretes of the series.

Table 7. Deicer salt scaling test results of field concretes after 50 cycles

Mixture	Visual rating	Mass of scaled particles (kg/m^2)
Durable field concretes — Series 1		
A10-01	0	0.08
A10-02	0	0.04
A10-03	0	0.06
A10-04	1	0.66
B10-05	0	0.04
C10-06	0	0.11
Non-durable field concretes — Series 2*		
N10-01	0	0.29
N10-02	0	0.46
N10-03	1.7	0.83
N10-04	1.7	1.08

Table 8. Deicer salt scaling test results of laboratory concretes after 50 cycles

Mixture	14 d. at 23° C		90 d. at 23° C		14 d. at 40° C	
	Visual rating	Scaled-off particles (kg/m^2)	Visual rating	Scaled-off particles (kg/m^2)	Visual rating	Scaled-off particles (kg/m^2)
D10-01	1.0	0.17	—	ND	3.7	2.42
D10-02	1.3	0.22	1.0	0.3	5	4.86
D10-03	1.7	0.60	2.7	1.0	5	10.49

The three laboratory concretes were subjected to three different drying procedures:

- normal drying for 14 days at 23° C and 50% R.H.;
- 90 days of drying under the same conditions;
- and 14 days of drying in an oven at 40° C.

For the specimens subjected to the standard ASTM C 672 drying period, the results in Table 8 show that the amount of residues after 50 cycles is always lower than the 1 kg/m² limit, though it clearly increases with the value of the air void spacing factor. It is interesting to note that, even with a 412 μm spacing factor, mixture D 10-03 still has an adequate scaling resistance (0.60 kg/m² after 50 cycles) according to the Swedish standard.

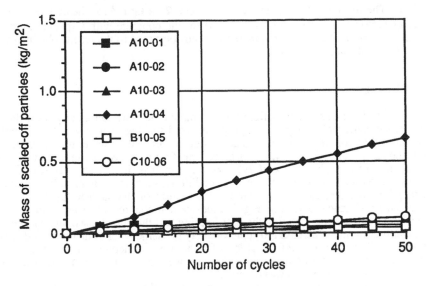

Fig. 1. Scaling behavior of the durable field concretes

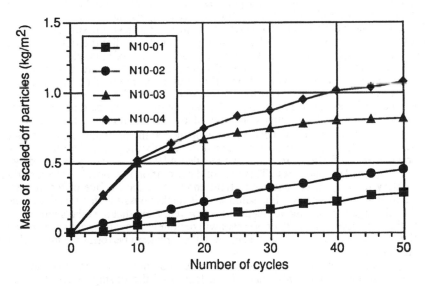

Fig. 2. Scaling behavior of the non-durable field concretes

After a 90 day period of drying under standard conditions, the results show that the resistance to scaling is lower, but all values are still equal to or below the 1 kg/m^2 limit. This confirms that the resistance to scaling determined in the laboratory tends to decrease as the length of the drying period before the tests increases [5, 6]. The results obtained with the specimens dried at 38°C for 14 days also confirm the results obtained by other investigators [5, 6]. Such harsh drying reduces considerably the resistance to scaling. The amount of residues varies between 2.42 kg/m^2 for the concrete with a spacing factor of 204 μm and 10.49 kg/m^2 for the concrete with a spacing factor of 412 μm, i.e. a more than ten fold increase when compared to the results obtained with the normally dried specimens.

5 Discussion

The results obtained from the series of non-durable field concretes clearly show that an adequate mechanical strength and a proper air void spacing factor are not sufficient to guarantee that concrete will be resistant to deicer salt scaling under natural exposure conditions. Concrete N 10-04, for instance, was found to have a strength of 42 MPa and an air void spacing factor of 192 μm. Nevertheless, it was severely scaled after only three winters of exposure. Various conditions, such as inadequate finishing and curing for instance, can weaken the microstructure of the surface layers and thus reduce the resistance to scaling. Fig. 2 indicates that concretes N 10-04 and N 10-03 were rapidly deteriorated during the first freezing and thawing cycles, but that the rate of deterioration then decreased to approximately the same level as that of the other two concretes. This tends to confirm, at least for these two concretes, the hypothesis of a weak surface layer, particularly since the tests were continued up to 100 cycles and little additional deterioration was observed.

Globally, the test results for the field concretes tend to show that the ASTM C 672 test is not too severe, since it allowed to distinguish quite clearly between the durable concretes (all but one with a very good laboratory scaling resistance) and the non-durable concretes (all with a lower laboratory scaling resistance). This would tend to show that the mechanisms of deterioration during the test are similar to the field mechanisms and that the ASTM test can thus allow to predict the field performance under natural exposure conditions with a reasonable degree of reliability.

The results of the laboratory test for mixture A 10-04 are unfortunately in contradiction with those obtained from all the other durable field concretes. Although mixture A 10-04 was found to be durable under natural exposure conditions, the amount of residues collected after 50 cycles was approximately ten times higher for this concrete than for all the other concretes of the series, i.e. 0.66 instead of 0.04 - 0.11 kg/m^2. This phenomenon can not be explained at the present time, but it could be related to the large amount of water added to the mixture just before the placement operations. It could also be related to the fact that A10-04 was mainly damaged by pop-outs during the laboratory test. This type of deterioration is probably enhanced during laboratory testing since, contrary to field conditions, the surface of the concrete is covered at all times with a saline solution. It is further possible that this field concrete will be damaged by scaling during the next few years, but that the others will not. Thus only the performance after 25 or 30 years will definitely allow to establish the reliability of the ASTM test, since it is the prediction of the long term performance that such a test should allow.

As previously mentioned, the results of the tests on the laboratory concretes clearly confirm the deleterious influence of the length of the drying period (and of the temperature during this period) on the resistance to scaling. Drying causes surface microcracking and, most important, can modify the porosity of the surface layers, particularly the continuity of the large capillaries [6-8]. This increases the freezable water content and thus reduces the scaling resistance [6, 9]. These results also indicate that drying at 40° C during 14 days is probably not a condition which is representative of field conditions. The amount of residues collected from the laboratory specimens dried at this temperature were found to vary between 2.42 and 10.49 kg/m^2, versus 0.04 and 1.08 kg/m^2 for the specimens from the field concretes.

From the results obtained from the field concretes tested, it would seem that the 1 kg/m^2 value that is commonly used to distinguish between good and poor scaling resistance is too high, since the amount of residues collected from most durable field concretes was found to be lower than 0.1 kg/m^2. Before reaching such a conclusion, however, it should be remembered that laboratory made concretes are tested after only 14 days of curing followed by 14 days of drying, whereas the durable field concretes tested were six years old. More data from systematic comparisons between laboratory and field performance are thus required before this limiting value can be redefined.

6 Conclusion

From the results of the laboratory tests that were performed on both field and laboratory concretes, it can be concluded that deicer salt scaling test procedures such as that described in ASTM C 672 are not too severe and allow to distinguish with a fair degree of reliability between those concretes which are scaling resistant and those which are not. The results further show that drying at 40° C during 14 days before the freezing and thawing cycles is most probably not a condition which is representative of field conditions, and that such a drying procedure should not be used before performing deicer salt scaling tests. More data from systematic comparisons between laboratory and field performance are required to better understand both the mechanisms of deterioration and the parameters that influence the microstructure of the surface layers. This should allow to define precisely the amount of residues after 50 cycles that represents an acceptable limit for good scaling resistance.

7 References

1.   SCHUBERT, P., LÜHR, H.P. (1976) Zur Prüfung von Betonwaren für den Straßenbau auf Frost und Tausalzwiderstand— Part I, *Concrete Precast Plant and Technology*, Vol. 42, N° 11, pp. 546-550, (in German).
2.   SCHUBERT, P., LÜHR, H.P. (1976b) Zur Prüfung von Betonwaren für den Straßenbau auf Frost und Tausalzwiderstand— Part I, *Concrete Precast Plant and Technology*, Vol. 42, N° 12, pp. 604-608, (in German).
3.   LITVAN, G.G., MACINNIS, C., GRATTAN-BELLEW, P.E. (1980) Cooperative test programme for precast concrete paving elements, in Durability of Building Materials and Components, <u>ASTM Special Technical Publication STP-691</u>, P.J. Sereda and G.G. Litvan Ed., pp. 560-573.
4.   SAUCIER, F., PIGEON, M., PLANTE, P. (1990) Field test of superplasticized concretes, *ACI Materials Journal*, Vol. 87, N° 1, pp. 3-11.
5.   SELLEVOLD, E.J., FARSTAD, T. (1991) Frost/salt testing of concrete: Effect of test parameters and concrete moisture history, *Nordic Concrete Research*, Vol. 10, pp.121-138.
6.   MARCHAND, J. (1993) Contribution à l'étude de la détérioration par écaillage du béton en présence de sels fondants, *Thèse de Doctorat*, École Nationale des Ponts et Chaussées, Paris, France, 316 p.
7.   FELDMAN, R.F. (1988) Effect of pre-drying on rate of water replacement from cement paste by propan-2-ol, *Il Cemento*, N° 3, pp. 193-201.
8.   PATEL, R.G., PARROTT, L.J., MARTIN, J.A., KILLOH, D.C. (1985) Gradients of microstructure and diffusion properties in cement paste caused by drying, *Cement and Concrete Research*, Vol. 15, N° 2, pp. 343-356.
9.   BAGER, D.H., SELLEVOLD, E.J. (1986) Ice formation in hardened cement paste - Part II: Drying and resaturation of room temperature cured paste, *Cement and Concrete Research*, Vol. 16, N° 6, pp. 835-844.

12

Scaling resistance tests of concrete – experience from practical use in Sweden

P.E. PETERSSON
Swedish National Testing and Research Institute, Boraas, Sweden

Abstract
A method for testing the scaling resistance of concrete, the "Boraas method" (SS 13 72 44), is presented. Using the method, the influence of the composition of ordinary Portland cements on the scaling resistance of concrete was investigated as well as the influence of different types and combinations of air entraining agents and plasticizers. Finally, results from practical experience of the method are presented. In Sweden the method is used, among other things, for pretesting concrete intended for bridges and also for continuous routine testing on specimens produced at the building site.
Keywords: Scaling resistance, frost resistance, test method, admixtures, cement, practical experience.

1 Introduction

Since 1988 the scaling resistance has been tested on all concrete used for bridges in Sweden. The tests are performed according to the Swedish standard test method SS 13 72 44 (the "Boraas" method).

Before the start of a building project the concrete must be specified and then approved in a pretesting procedure where four specimens are tested. During the building process specimens are produced at the building site for continuous routine testing. So far all the tests have been performed at the Swedish National Testing and Research Institute in Boraas.

Since the introduction of the obligatory scaling resistance test for bridges in Sweden the quality of the concrete has improved considerably, at least where the scaling resistance is concerned. The main reason for this is probably that people involved in the production and handling of concrete have become more conscious about the importance of "quality thinking". They also know that every mistake will be detected by the testing procedure. Today it therefore is unusual to have too low air content or too high water-cement ratios etc. when the concrete is delivered at the building site.

Using the test method it has been shown that poor combinations of admixtures may ruin durability, and also that the choice of cement is important for achieving good scaling resistance. Different ordinary Portland cements may also result in

Freeze-Thaw Durability of Concrete. Edited by J. Marchand, M. Pigeon and M. Setzer.
Published in 1997 by E & FN Spon, 2–6 Boundary Row, London SE1 8HN, UK.
ISBN 0 419 20000 2.

completely different scaling values. These findings have been adopted by the concrete producers in Sweden, which has resulted in improved concrete quality.

2   Test procedure

In the standard procedure according to SS 13 72 44, 50 mm thick slices are sawn from 150 mm cubes. The saw cut is placed in the centre of the cube and perpendicular to the top surface (i.e. the cast surface) of the cube. After casting, the cubes are stored in water (+20°C) for seven days and then in the air (+20°C, RH=50%, wind velocity less than 0.1 m/s) until day 21 when the sawing takes place. Then the specimens are placed in the same climate again until day 28. During this period rubber cloth is glued to all the sides but the test surface. The rubber cloth reaches 20 mm above the test surface and this rubber edge makes it possible to keep a salt solution on the test surface during the test. The testing arrangement is shown in Figure 1.

On day 28 water is poured on to the test surface. The resaturation lasts for three days. Then the water is replaced with a 3 mm thick layer of 3% sodium-chloride solution. The salt solution is covered with a plastic film in order to prevent evaporation which may influence the salt concentration and thereby also the test result as reported by Verbeck and Klieger (1956). Finally all sides but the test surface are covered with a 20 mm thick layer of heat insulation which makes the heat transport one-dimensional and makes the cooling of the specimen take place through the test surface and not through the bottom of the specimen.

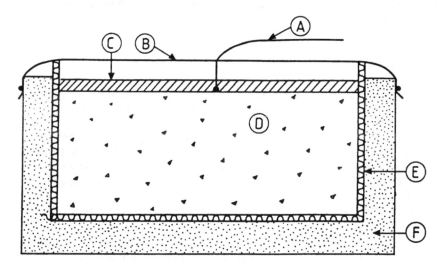

Figure 1.   The testing equipment used for the freeze-thaw test according to SS 13 72 44 (the Boraas method). A=temperature measuring device, B=plastic film, C=salt solution, D=specimen, E=rubber cloth, F=heat insulation.

The specimens are then placed in a freezer and subjected to freeze-thaw cycles according to Figure 2. The temperature is registered in the salt solution for at least one specimen in each freezer. After 7, 14, 28, 42 and 56 cycles the scaled material is collected and dried at +105° C, after which the weight is measured. The results are given as the amount of scaled material per unit area.

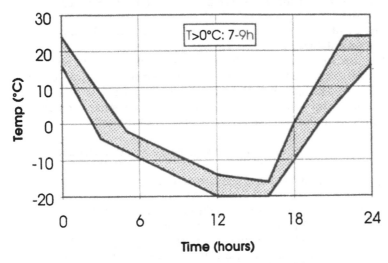

Figure 2.   The temperature cycle used for SS 13 72 44

The scaling resistance of the concrete can be assessed according to Table 1. Normally four specimens are tested. The table is primarily applicable to normal OPC concrete. For other concrete types (for example, vacuum-treated concrete, silica fume concrete, concrete with new types of admixtures, etc.) it may sometimes be necessary to use a somewhat different assessment procedure, for example material scaled after a larger number of freezing cycles, see for example Petersson (1986).

Table 1.   Assessment of concrete tested according to the Swedish standard test method SS 13 72 44.

Scaling resistance	Requirement
Very good	No specimen has more than 0.1 kg/m^2 scaled after 56 cycles.
Good	The mean value for the material scaled after 56 cycles (m_{56}) is less than 0.5 kg/m^2, and m_{56}/m_{28} is less than 2
Acceptable	The mean value for the material scaled after 56 cycles (m_{56}) is less than 1.0 kg/m^2 and m_{56}/m_{28} is less than 2.
Unacceptable	The requirements for acceptable scaling resistance are not met

3   Factors influencing the scaling resistance of concrete

3.1   Introduction

In addition to the air content and the water-cement ratio there are a number of other parameters which may influence the scaling resistance of concrete. Using the test method described above, the influence of two such parameters has been investigated: the type and combinations of admixtures and the composition of the cement.

3.2   The influence of the type of and combination of air-entraining and plasticizing admixtures on the scaling resistance

Petersson (1986) compared a number of different concrete mixtures with respect to their ability to resist a combined attack of frost and de-icing agents. Two air-entraining agents, AEA1 and AEA2, were used as well as a combination of an AEA and a plasticizer. The admixtures were added to the fresh concrete according to the producer´s recommendations. The admixtures were of the following types:

> AEA1:   a mixture of neutralized Vinsol resin+synthetic tenside
> AEA2:   a neutralized Vinsol resin
> PL: a melamine-based plasticizing agent

A Swedish ordinary Portland cement (Degerhamn anl, SR) was used with a composition according to Table 2 in section 3.3. The water-cement ratio (W/C) was 0.43-0.47 and the slump 75-85 mm for all the batches. The aggregate used was a natural gravel with a maximum particle size of 16 mm. Concrete mixtures with air contents of 3, 4, 5 and 6.2% by volume were produced, and a deviation of 0.2% from the nominal value was accepted.

The results are presented in Figure 3. For the concrete with AEA1 as the only admixture, the amount of scaled material is small for air contents exceeding about 4% but even when the air content is as low as 3% the concrete has an acceptable salt frost resistance according to SS 13 72 44. For the concrete with AEA2 the scaling resistance is also good for air contents exceeding about 4% while the salt-frost resistance seems to be poor for an air content of 3%.

The scaling values are high for the combination of air entraining agent and plasticizer used. Not even when the air content is as high as 6% is the requirement for an acceptable salt-frost resistance according to SS 13 72 44 fulfilled.

The results indicate that a combination of an air entraining agent and a plasticizer gives much poorer results than when the only admixture is an AEA. For a good AEA it seems to be possible to use an air content 2-3% lower than for a combination of admixtures and still obtain the same resistance to frost damage in a saline environment. Other combinations may result in other results but one must always be aware of the difficulties that may arise when combinations of AEAs and plasticizers are used.

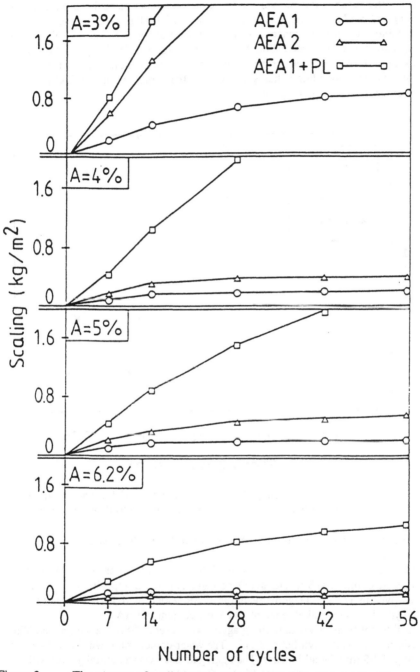

Figure 3.   The amount of scaling vs the number of freeze-thaw-cycles for different air contents and different types and combinations of admixtures (according to Petersson 1989).

In Figure 4 the scaling after 56 cycles is presented as a function of the spacing factor. According to SS 13 72 44 the scaling must not exceed 1.0 kg/m^2 for a concrete with an acceptable salt-frost resistance. In order to obtain this requirement the value of the spacing factor must not exceed about 0.20 mm, see Figure 4. With a good AEA as the only admixture, this value can be reached for an air content not higher than 3-4%, while the corresponding value is over 6% for the combination of AEA and plasticizer used in this investigation.

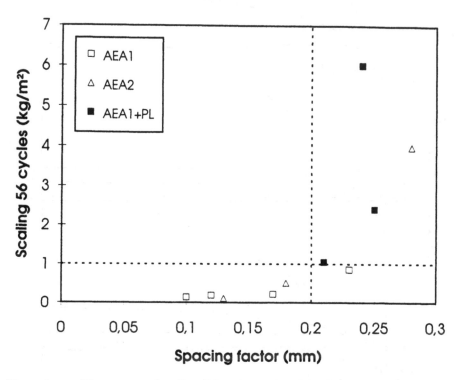

Figure 4.   The amount of scaling (56 cycles) as function of the spacing factor for the results presented in Figure 3 (according to Petersson 1989).

3.3 The influence of the cement composition on the scaling resistance of concrete

In an investigation reported by Malmström (1990) two Swedish ordinary Portland cements were compared; Slite std and Degerhamn anl. The composition of the cements is given in Table 2. Degerhamn anl is an SR cement with a low value of the heat of hydration, a low alkali content and a low content of C$_3$A.

Two different types of air-entraining agents were used. C88L is a neutralized Vinsol Resin and Aer L is a synthetic tenside. The water-cement ratio was 0.43-0.46 and the slump 75-85 mm for all batches. The aggregate used was a natural gravel with a maximum aggregate particle size of 16 mm. Concrete mixtures with air contents ranging from 2 to 6% by volume were used.

Table 2.   The composition of the cements used in the investigation

Component	Slite std % by weight	Degerhamn anl % by weight
SiO_2	20.4	22.6
TiO_2	0.24	0.20
Fe_2O_3	2.16	4.29
Al_2O_3	4.58	3.41
MnO	0.06	0.22
CaO	63.9	66.3
MgO	3.16	1.19
Na_2O	0.25	0.13
K_2O	1.28	0.70
SO_3	3.4	2.4
C_3S	61.2	62.4
C_2S	12.3	17.7
C_3A	8.5	1.8
C_4AF	6.6	13.1
Heat of hydr (7 days)	260 kJ/kg	320 kJ/kg

The results are presented in Figure 5, where the scaling after 56 cycles is shown as a function of the air content. According to the test results the cement appears to have a marked influence on the scaling resistance, while the two types of AEA produce about the same results. For the cement with high alkali (Slite std), the air content must exceed about 5-6% if acceptable scaling resistance according to SS 13 72 44 is to be obtained. The corresponding value for the low-alkali-cement is about 3-4%. This means that the choice of cement may influence the required air content, even for different types of Portland cement.

To date there are no definitive explanation for the difference between the cements but the alkali content and perhaps also the C3A content probably influence the scaling resistance.

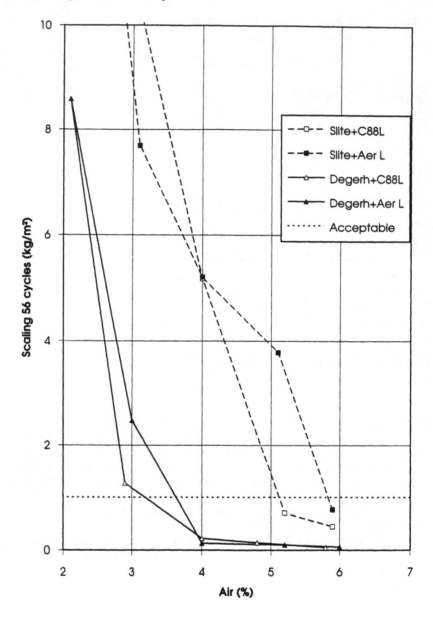

Figure 5.   Scaling after 56 cycles as function of air content for two ordinary
             Portland cements and two air entraining admixtures (according to
             Malmström 1990).

4   Experience from practical use of the Boraas-method in Sweden

All concrete used for bridges in Sweden has to be tested for scaling resistance according to the Boraas method. The testing includes two phases:

1. Before the start of the construction the actual concrete mix must be pretested and must fulfil the requirements for good scaling resistance, i.e. the scaling must not exceed 0.5 kg/m^2 after 56 cycles. The tests are performed on series of four specimens.

2. During construction the concrete is subjected to continuous routine testing. A specified number of specimens, depending on the size of the bridge, are produced at the building site, and the total number of specimens must not be less than three for each bridge. The requirement for this continuous testing is that the scaling must not exceed 1.0 kg/m^2 after 56 cycles, i.e. the scaling resistance must be rated as acceptable.

Since the scaling resistance test became obligatory, the quality of the concrete for Swedish bridges has improved considerably. As mentioned above this depends, among other things on the fact that people involved in the production of concrete have improved their "quality thinking". For example, the concrete delivered at the building site today almost always fulfils or is better than the requirements of the order specifications. It also means that the materials used for the production of concrete is with few exceptions, of high quality. The low alkali cement called Degerhamn anl, see Figure and Table 2, is used for **all** concrete in the Swedish bridges and the admixtures or combinations of admixtures used are always tested to see that they perform well in the actual concrete mix.

All the 240 pretesting results carried out at the Swedish National Testing and Research Institute in Boraas in 1990 are presented, in chronological order, in Figure 6. Each dot represents the mean value of four specimens. Only six results (2.5 %) do not fulfil the requirements at pretesting, i.e. 0.5 kg/m^2 after 56 cycles, while the scaling for 220 series (92 %) is less than 0.2 kg/m^2. The mean value for all the results are 0.10 kg/m^2.

In Figure 7, the 165 results from the continuous routine testing during the first half year of 1990 are presented. Each dot represents a single specimen. Only one specimen (0.6%) does not fulfil the requirement for continuous testing which is 1.0 kg/m^2 after 56 cycles and 94 % of the results are below 0.2 kg/m^2. The mean value is 0.09 kg/m^2.

The results indicate that the concrete used for bridges in Sweden is of high quality, at least where the scaling resistance is concerned. It can also be seen that there is a good correlation between the pretesting results and the results from the continuous testing. The pretesting therefore seems be useful for predicting the scaling properties for the concrete used at the building site.

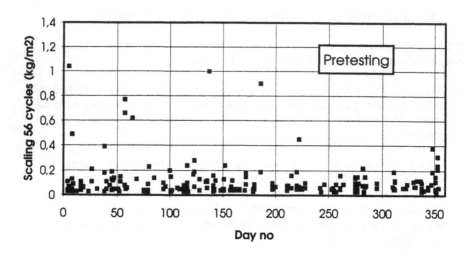

Figure 6.   Pretesting results of the scaling resistance of concrete carried out at the Swedish National Testing and Research Institute during 1990.

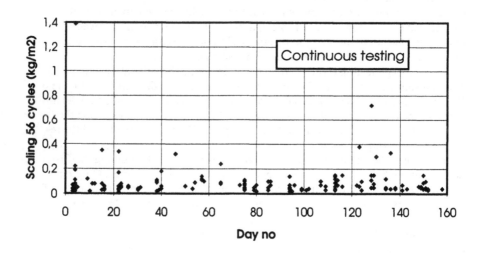

Figure 7.   Scaling resistance results from continuous routine testing during the first six months of 1990. The tests were performed at the Swedish National Testing and Research Institute on specimens produced at building sites.

5   Summary and conclusions

A test method for determining the scaling resistance of concrete, the "Boraas" method (SS 13 72 44) is described in the paper. Using the method it was found that concrete with an air-entraining agent as the only admixture normally gives better scaling resistance than when the AEA is combined with a plasticizer. It was also found that the composition of ordinary Portland cements has a marked effect on the scaling resistance. In this investigation a SR cement with a low alkali content (Degerhamn anl) produced a much better scaling resistance than when another Swedish cement with a high alkali content (Slite std) was used.

Results from practical experience of the test method in Sweden are presented. Pretesting results relevant to concrete for bridges are in good correlation with results from continuous routine testing on specimens produced at the building sites. The results also indicate that the concrete used for bridges in Sweden is normally of high quality, at least where scaling resistance is concerned.

6   References

Malmström, K (1990). The importance of cement composition on the salt-frost resistance of concrete. Technical report 1990:07, Swedish National Testing Institute, Boraas, Sweden (in Swedish).

Petersson P.E. (1986). The influence of silica fume on the salt frost resistance of concrete. Technical report SP-RAPP 1986:32, Swedish National Testing Institute, Boraas, Sweden.

Petersson P.E. (1989). The use of air-entraining and plasticizing admixtures for producing concrete with good salt-frost resistance. Technical report SP-RAPP 1989:37, Swedish National Testing Institute, Boraas, Sweden.

SS 13 72 44 (1988). Concrete testing-Hardened concrete-Frost resistance. Swedish standard test method. Swedish standards institution, Stockholm, Sweden.

Verbeck, G.J. and Klieger, P. (1957) Studies of salt scaling of concrete. Highway research board, Bulletin 150.

13

Round robin tests on concrete frost resistance

H. KUKKO AND H. PAROLL
Technical Research Centre of Finland, Building Materials Laboratory, Espoo, Finland

Abstract

Six Nordic research institutes participated in this Round Robin project financed by NORDTEST. The test methods used in the Round Robin tests were:

- SS 13 72 44, Procedure 1, Method A
- Critical degree of saturation method S_{cr} (modified)
- ASTM C 666, Method A (modified)
- Norwegian PF - method, which is closely related to the Finnish SFS 4475 "Protective pore ratio"-method.

The frost resistance assessments made by the participating laboratories with SS 13 72 44 were in good agreement. No reliable results were obtained with the critical degree of saturation method. The main reason seems to be the uncontrollable cracking of the specimens during the test. Only two institutes carried out the ASTM C 666 test. The results from the three participants in the PF - method comparison are in good agreement. The method is easy to carry out and can be used for quality control, but is less known internationally.

Keywords: Concrete, frost, frost resistance, salt, testing, assessment, Round Robin

1 Background and purpose of the tests

In this study, some methods used in the Nordic countries were compared. The reason for this comparison is the work done in CEN (European Committee for Standardization) for common European standards. The purpose was to clarify the applicability of Nordic test methods for more general use.

Freeze-Thaw Durability of Concrete. Edited by J. Marchand, M. Pigeon and M. Setzer.
Published in 1997 by E & FN Spon, 2–6 Boundary Row, London SE1 8HN, UK.
ISBN 0 419 20000 2.

In this Round Robin test, four different frost resistance testing methods were chosen. Six Nordic research institutes participated in the project. The participants are presented in Table 1.

Table 1. The participating institutes

Abbreviation	Institute
NBI	Norwegian Building Research Institute P.O.box 123, Blindern 0314 Oslo, Norway
RB	Icelandic Building Research Institute Rb-Keldnaholti IS-112 Reykjavik, Iceland
SBI	Danish Building Research Institute P.O.box 119 DK-2970 Hörsholm, Denmark
SIB	The National Swedish Institute for Building Research Box 758, 80129 Gävle, Sweden
SP	Swedish National Testing and Research Institute P.O.box 857, S-50115 Borås, Sweden
VTT	Technical Research Centre of Finland P.O.box 26, 02150 Espoo, Finland

The four following methods were chosen to test the frost resistance of concrete in the Round Robin test:

1. SS 13 72 44, Procedure 1, Method A
   This method is based on repeated freezing and thawing of concrete covered with a 3% NaCl solution. The test specimens are heat insulated from all sides except sawn upper surface, which is first immersed in pure water. Immediately before the freeze-thaw cycling, the water on the specimen is exchanged for 3% NaCl solution. The freezing and thawing takes place in air (1).
2. Critical degree of saturation method S_{cr} (modified)
   This method, developed by Prof. Fagerlund in Sweden, is based on the existence of critical degrees of saturation at freezing. The degree of saturation is defined as the ratio of evaporable water (105 °C) to the total open pore volume. The method follows in all essential parts the RILEM tentative recommendation (2). In practice, the test is divided into two parts:
   1. a freeze-thaw test to determine the critical degree of saturation and
   2. a capillary water uptake test to determine pore volumes and water uptake rates.
3. ASTM C 666, Method A (modified)
   This method includes freezing and thawing of concrete specimens, both taking place in water. In this method the thawing temperature is essentially lower (4.4 °C) than in the two first methods, where it is around 20 °C (3).

4. PF - method (SFS 4475 "Protective pore ratio" modified to be equivalent to the Norwegian PF - method)

This method is based on the idea of determining two kinds of pores in hardened concrete: Cement paste pores which fill by suction when a piece of concrete is submerged in water; and macro/air pores which only fill under high water pressure. The "protective pore ratio" is the ratio of protective air and macro pores to the total porosity. It is often assumed that this ratio should be higher than the freezing expansion of water, 0.09, with a safety margin (4).

In the Round Robin test, two qualities of concrete were used:

- concrete A, compressive strength 30 MPa with an air-entraining agent
- concrete B, compressive strength 60 MPa without an air-entraining agent.

Concrete A is a concrete commonly used in Sweden when good frost resistance is required. Concrete B is a non-air-entrained concrete of relatively high strength. In spite of its strength it often has a tendency to crack extensively in frost tests. The samples of concrete A were cast on 10.9. 1992 by the Swedish National Testing and Research Institute (SP) and the samples of concrete B were cast on 10.9. 1992 by the Technical Research Centre of Finland (VTT). SP and VTT sent the samples to the participating institutions. The mix designs are presented in Table 2.

Table 2. Mix designs for concretes A (SP) and B (VTT).

		Concrete A (SP)		Concrete B (VTT)	
Max. particle size	mm		16		16
Cement	kg/m³	Degerhamn Std P	357	Partek P 40/28	387
Aggregate - fraction	kg/m³	0 - 0.125 mm	53	0 - 0.125 mm	115
	kg/m³	0.125 - 0.25 mm	88	0.1 - 0.6 mm	154
	kg/m³	0.25 - 0.5 mm	230	0.5 - 1.2 mm	154
	kg/m³	0.5 - 1 mm	159	1 - 2 mm	307
	kg/m³	1 - 2 mm	212	2 - 3 mm	154
	kg/m³	2 - 4 mm	177	3 - 5 mm	192
	kg/m³	4 - 8 mm	212	5 - 10 mm	365
	kg/m³	8 - 16 mm	638	8 - 16 mm	480
Additives		air-entraining		plasticizer	
	kg/m³	Cementa L 14	0.09	Scancem SP 62	7.75
Water	kg/m³		175		135
Water cement ratio	w/c		0.49		0.35
Aggregate/cement ratio			4.96		4.96

Table 3. Properties of fresh concrete.

		Concrete A (SP)	Concrete B (VTT)
Consistency			
- slump	cm	8.0	7.5
- vebe consistency time	sVB	-	2.9
- air content	%	5.2	2.7
Density	kg/m^3	2300	2450

Three participating institutes determined the compressive strength for the concretes A and B at an age of 28 d. The results are presented in Table 5.

The tests carried out and the number of the specimens to be tested by the institutes are presented in Table 4.

Table 4. Number of specimens and different tests. Concrete A: 30 MPa, with air-entraining agent (SP). Concrete B: 60 MPa, with air-entraining agent (VTT)

Test	Sample	NBI	RB	SBI	SIB	SP	VTT
SS 13 72 44	Cube 150 mm	3A+3B	3A+3B	3A+3B	3A+3B	3A+3B	3A+3B
S_{cr}	Beam 100x100x500 mm	-	-	-	3A+3B 1)	3A+3B	3A+3B
ASTM C 666 method A	Beam 100x100x500 mm	3A+3B	3A+3B	-	-	-	-
PF - method (SFS 4475)	Cube 100 mm	3A+3B	-	-	-	3A+3B	3A+3B

1) Testing unsuccessful. No results reported.

2 Test results

2.1 Compressive strength of concretes A and B

Three of the participating institutes determined the compressive strength at 28 d. The results are presented in Table 5. The size of the specimens was 100 mm x 100 mm x 100 mm.

Table 5. Compressive strength at an age of 28 d.

Institute		NBI	SP	VTT
Concrete A (SP)	MPa	46.5	43.4	42.3
Concrete B (VTT)	MPa	72.8	70.0	69.0

2.2 SS 13 72 44, Procedure 1, Method A

A summary of the results obtained by the participants for concretes A and B is presented in Tables 6 and 7.

Table 6. Test results. SS 13 72 44, Procedure 1, Method A. Concrete A (SP).

Institute	Scaled material 56 d average, kg/m^2	Variation coefficient %	Ratio m_{56}/m_{28} average	Assessment of frost resistance
NBI	0.01	73	1	Very good
RB	0.3	44	1.6	Good
SBI	0.13	129	1.9	Good
SIB	0.06	31	1.3	Very good
SP	0.15	28	1.4	Good
VTT	0.13	46	1.4	Good

Table 7. Test results. SS 137244, SS 13 72 44, Procedure 1, Method A. Concrete B (VTT).

Institute	Scaled material 56 d average, kg/m^2	Variation coefficient %	Ratio m_{56}/m_{28} average	Assessment of frost resistance
NBI	1.24	56	1.6	Unacceptable
RB	3.5	9	1.8	Unacceptable
SBI	3.26	16	2.4	Unacceptable
SIB	2.95	12	2.1	Unacceptable
SP	4.25	19	2.0	Unacceptable
VTT	2.49	40	2.3	Unacceptable

2.3 Critical degree of saturation method S_{cr} (modified).

This test was found out to be difficult to perform. SIB reported that their test was unsuccessful and no results were returned. VTT and SP reported that the pressure test caused cracks in the tested samples of concrete A. With samples of concrete B, VTT determined the critical degree of saturation S_{cr} to 0.90 with a large dispersion (Fig. 1).

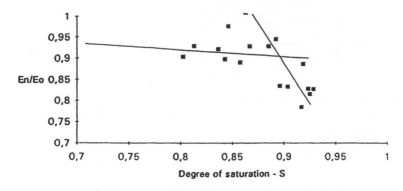

Fig. 1. Determination of S_{cr} (dynamic E-modulus is used as the criterion).

2.4 ASTM C 666 method A (modified)

In this frost resistance test the durability factor DF was determined. The results are presented in Table 8.

Table 8. Determination of the durability factor DF.

Institute	Concrete A (SP) DF	Concrete B (VTT) DF
NBI	91	9
RB	93	_ 1)

[1] Relative dynamic E - modulus < 40 % after 73 cycles.

2.5 PF-method (SFS 4475 modified)

In the PF-method the pore protection factor PF was determined for the assessment of the frost resistance of concrete. The results are presented in Table 9.

Table 9. The determined PF-values for the assessment of the frost resistance of concrete.

Institute	Concrete A (SP) PF (%)	Assessment of frost resistance	Concrete B (VTT) PF (%)	Assessment of frost resistance
NBI	23	Frost resistant	19.9	Not frost resistant
SP	27	Frost resistant	17	Not frost resistant
VTT	29	Frost resistant	18	Not frost resistant

3 Discussion of the test results

3.1 General
In this Round Robin test the comparison between the methods of assessing of the frost resistance of concrete could be made with a frost resistant and a frost irresistant concrete. The reliability of the methods can be judged by comparing the results of the participants. The question of how easily the tests can be carried out in practice was also discussed.

3.2. SS 13 72 44, Procedure 1, Method A
This test was carried out by all the participating institutes. The frost resistance assessments made by the participating institutes were in good agreement.

The frost resistance of concrete A was assessed as good by four institutes and very good by two institutes. For the good assessment the average of the scaled material at an age of 56 d varies between 0.13 - 0.3 kg/m^3 and the average for the ratio m_{56}/m_{28} between 1.4 - 1.9.

The frost resistance of concrete B was assessed as unacceptable by all six institutes. For the unacceptable assessment the average of the scaled material at an age of 56 d varies between 1.24 - 4.25 kg/m^3 and the average for the ratio m_{56}/m_{28} between 1.6 - 2.4. This greater range is not unusual for frost irresistant concrete.

The good agreement of the frost resistance assessments made by the participating institutes shows that this method is reliable.

This method also provides criteria for the assessment of the frost resistance of concrete: very good, good, acceptable and unacceptable. This fact increases the applicability of the method.

3.3 Critical degree of saturation method S_{cr} (modified)
No reliable results were obtained with this method. The main reason seems to be the uncontrollable cracking of the specimens during the test.

3.4 ASTM C 666, Method A (modified)
Only two institutes (NBI and RB) carried out this test. The reason for the scant participation is variation among the direct freeze-thaw procedures used in the Nordic countries. Some institutes use a method resembling ASTM 666 procedure B, and therefore did not participate in this part of the tests.

The numerical values for the durability factor (DF) are in agreement for concrete A and concrete B (Table 8). However, the standard does not give any criteria for the assessment of the frost resistance of concrete based on the numerical results of the test. This is a shortcoming because for the reliable assessment of the frost resistance of concrete one has to compare the results of this method to assessments based on the results of the SS 13 72 44 and/or PF - methods.

3.5 PF - method (SFS 4475 "Protective pore ratio" modified to be equivalent to the Norwegian PF - method)

Three institutes (NBI, SP and VTT) carried out this test. The method is only used in Finland and Norway.

The numerical values for the durability factor (PF) are in agreement for concrete A and concrete B (Table 9).

The method description does not give any criteria for the assessment of the frost resistance of concrete based on the numerical results (PF-values) of the test. In practice the assessment is based on the criteria given in the Finnish standard SFS 4475 "Protective pore ratio" which gives a value of 0.20 for normal cases, e.g. facades. According to this standard one can assess that concrete A is frost resistant and concrete B is not frost resistant. The assessment of all three participants are in agreement. The method is easy to carry out.

4 Conclusions

In this project the participants have carried out the Round Robin test with four methods to assess the frost resistance of concrete. When examining the assessments, attention has been paid to the reliability and precision of the results and to how difficult the tests were to carry out in practice.

Considering the above-mentioned factors, the effects of which have became apparent during the test, the testing of the frost resistance of concrete according to the standard SS 13 72 44 "Concrete testing - Hardened concrete - Frost resistance" is considered by the participants to be the most suitable among those used in Nordic countries as a CEN - standard. The PF - method (SFS 4475 "Protective pore ratio" modified to be equivalent to the Norwegian PF - method) can be used for quality control but is less known internationally.

5 Literature

1. Swedish Standard SS 13 72 44 Concrete testing - Hardenend concrete - Frost resistance, English transl. Swedish Testing and Research Institute 1992 (similar to ASTM C672).
2. Fagerlund, G. The critical degree of saturation method of assessing the freeze-thaw resistance of concrete.   Prepared on behalf of RILEM Committee 4 CDC. Tentative recommendation. Materials and Structures 1977:58. pp. 217 - 229.
3. ASTM C666 Standard test method for frost resistant concrete to rapid freezing and thawing.
4. Sellevold E.J. Hardened Concrete - Determination of air/macro and gel/capillary porosity (PF - method), NBI report O 1731, The Norwegian Building Research Institute 1986 (In Norwegian with appendix draft method in English).

14

Three different methods for testing the freeze-thaw resistance of concrete with and without de-icing salt

E. SIEBEL AND T. RESCHKE
Research Institute for the Cement Industry, Düsseldorf, Germany

Abstract
In the various countries in Europe there are large numbers of test methods for assessing the durability of a concrete with respect to its freeze-thaw resistance. In order to investigate the suitability and reproducibility of different national test methods the CEN working group TC 51/WG 12/TG 4 "Freeze-thaw Resistance" decided to carry out a European Round Robin test. Three of the more frequently used methods - the Scandinavian slab method, the German cube method and the CDF-test - were selected for the Round Robin test. With all three of the methods used it was possible to differentiate between a concrete with high freeze-thaw resistance with de-icing salt and a concrete with inadequate resistance. While testing without de-icing salt it was somewhat better to differentiate between the concretes with the cube and CDF methods than with the slab method. All three methods have to be made more precise in order to reduce the comparative variations.
Keywords: European standardization, freeze-thaw-resistance, test methods.

1 Introduction

Not only is the load bearing capacity of a structure of

Freeze-Thaw Durability of Concrete. Edited by J. Marchand, M. Pigeon and M. Setzer.
Published in 1997 by E & FN Spon, 2–6 Boundary Row, London SE1 8HN, UK.
ISBN 0 419 20000 2.

great importance, but also its durability, and therefore the building materials which are used in its construction. If structures and their components are exposed directly to the effects of weather then, to be durable, they must have an adequate resistance to freezing and thawing, and in certain cases - chiefly in road construction - to freezing and thawing in the presence of de-icing salt, throughout their useful lifes. This requires compliance with the requirements laid down for basic materials and for concrete composition and production, which are specified in the standards and regulations (description concept). When new basic materials or untried compositions and methods of production are used it must be possible to test the required properties, e.g. high freeze-thaw resistance with and without de-icing salt. This also applies to the testing of ready-mixed concrete, of concrete products and precast concrete members [1].

Not all the great number of parameters which influence the freeze-thaw resistance of concrete with and without de-icing salt can be measured quantitatively. The freeze-thaw resistance can therefore only be assessed by using a test method which covers the complete damage mechanism as a whole by imitating practical conditions. Any such method represents a convention, and the results which it produces are not absolute values but enable a relative evaluation to be made.

In the various countries in Europe there are large numbers of test methods for assessing the durability of a concrete with respect to its freeze-thaw resistance. In the context of European standardization it is therefore necessary to select suitable test methods and draw up a European standard. The CEN working group TC 51/WG 12/TG 4 "Freeze-thaw Resistance" has the task of drawing up such a standard. In order to investigate the suitability and reproducibility of different national test methods it was decided to carry out a European Round Robin test. Three of the more frequently used methods were selected for the test.

Some important requirements for these test methods were that they should provide a sufficiently sharp differentiation to permit classification into low, medium or high resistance, that they should be simple to perform, and that they should measure damage resulting both from scaling and from deeper-reaching destruction of the internal structure of the concrete.

2 Test methods

When the above-mentioned requirements were taken into account the suitable methods turned out to be the Scandinavian slab method [2] which is used frequently in some Northern European countries, and the German cube method [3], for which a great deal of experience is available in Germany. The CDF method [4], which was originally intended only for testing the resistance to frost in combination with de-icing salt, was also proposed by the Rilem TC 117.

2.1 The Scandinavian slab method

The Scandinavian slab method - a further development of the Austrian method - is described in detail in the Swedish standard SS137244 "Concrete testing - Hardened concrete - Frost resistance". It can be used to test freeze-thaw resistance with water, or with a 3% NaCl solution. The particular test liquid stands on the surface of the test piece to a depth of about 3 mm. Apart from the test surfaces the test pieces are sealed with rubber. The bottom and side surfaces are also thermally insulated and the surface is covered with plastic film (Fig. 1).

The specimens which have been subjected to freezing are measured after 7, 14, 28, 42 and 56 freeze-thaw cycles, and the scaling from the test surface is specified in kg/m^2. The testing takes place in a frost chamber. One freeze-thaw cycle covers 24 hours, and the temperature in the concrete ranges between +20 ± 4°C and -18 ± 2°C.

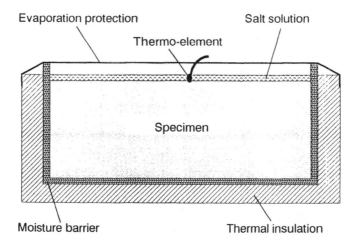

Fig. 1 Scandinavian Slab Test

The test pieces are slabs with dimensions 15 x 15 x 5 cm sawn from cubes with an edge length of 15 cm.

After they have been made the specially prepared cubes are stored as follows: 1 day in the mould, 6 days under water (20 ± 2°C), then storage in a climatic chamber at 20 ± 2°C days and 50 ± 10% relative humidity. At the age of 21 ± 2 days the slabs are sawn, provided with the sealing and insulation, and stored for the following 7 days in the climatic chamber at 20°C and 50% relative humidity. The test liquid is poured in after these 7 days, and the freeze-thaw cycles then start after 3 days.

The method has now been developed further so that it is also possible to determine internal damage to the structure by measuring the acoustic transit time.

2.2   German cube method
Cubes of 10 cm edge length which are made up specially or taken from components or concrete ware are frozen in a container filled with water (Fig. 2) and thawed again. One freeze-thaw cycle covers 24 hours. To test the freeze-thaw resistance with de-icing salt, a 3% NaCl solution is used instead of water. The loss in weight is measured by weighing the weathered components after 7, 14, 28, 42 and 56 freeze-thaw cycles, and is specified in wt.%. The testing is carried out in a frost chest, and the system is thawed by flooding the chest with water. The temperature in the concrete ranges between +20 ± 2°C and -15 ± 2°C.

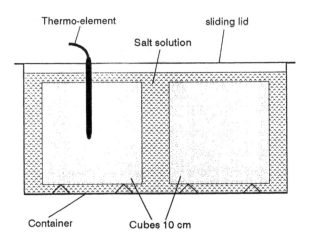

Fig. 2   Container with cubes and temperature sensor

After they have been made the specially prepared cubes are stored as follows: 1 day in the mould, 6 days under water (20 ± 2)°C, 20 days in a 20/65 climatic chamber and 1 day in the test liquid (water or 3% NaCl solution). The testing therefore starts on the 28th day.

The method has now been developed further so that it is also possible to measure damage to the internal structure. The dynamic elastic modulus and its change through exposure to frost are assessed by measuring the acoustic transit time.

2.3 CDF-Test

The CDF-test ("Capillary Suction of De-icing Chemicals and Freeze-Thaw-Test") was developed for testing the freeze-thaw resistance of concrete with de-icing salt. The samples are stored for 21 days at 20°C and 65% relative humidity and then placed in a sample container for 7 days in a foot bath (water or 3% NaCl solution). During this time a certain equilibrium moisture level is reached through "capillary suction". The exposure to freeze-thaw cycles is then carried out in 12 hour cycles. 28 freeze-thaw cycles are sufficient for control testing. The weathering losses are measured after 4, 8, 16 and 28 freeze-thaw cycles. The testing takes place in a liquid-cooled climatic chest in which the temperature cycle can be maintained very accurately (Fig. 3). The test temperature ranges between +20°C and -20°C.

The test pieces are slabs with the dimensions 15 x 15 x 7.5 cm. After they have been made the specially prepared

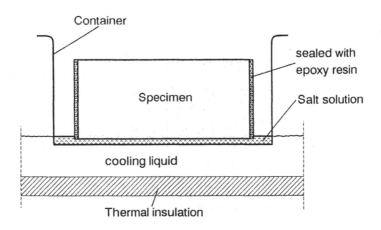

Fig. 3 CDF-Test

samples are stored as follows: 1 day in the mould, 6 days under water, then in the climatic chamber at 20°C and 65% relative humidity until 28 days. Before the start of the capillary suction the side surfaces of the dry samples are sealed with a solvent-free epoxy resin.

3 Producing the concrete

In order to keep the variation in concrete quality of the test pieces as low as possible all the samples were produced centrally by the laboratory of the Research Institute in Düsseldorf, and then sent to the testing institutes taking part in the Round Robin test.

Three different concretes were produced (see Table 1). Concretes 1 and 2A were tested for the freeze-thaw resistance with de-icing salt, and concretes 2B and 3 were tested for freeze-thaw resistance without de-icing salt. The properties of the unset and hardened concretes are shown in Table 2.

Table 1 Mix-design

MIX-Nr.	MIX-Design	freeze-thaw test	
		with de-icing chemicals	without de-icing chemicals
1	$c = 350$ kg/m^3 w/c = 0,45 air content 5...6 %	X	
2 A	$c = 350$ kg/m^3 w/c = 0,45	X	
2 B	no air-entrainment		X
3	$c = 280$ kg/m^3 w/c = 0,65 no air-entrainment		X

cement   :   CEM I 32.5 R
aggregates: natural sand and gravel, maximum size: 16 mm

Table 2   Properties of concrete

Mix-No.	MIX 1	MIX 2 A	MIX 2 B	MIX 3
1.) Fresh concrete :				
consistency	1.37 [1]	1.46 [1]	1.44 [1]	42 [2]
bulk density in kg/dm^3	2.29	2.38	2.38	2.34
air content in %	5.1	2.2	2.2	2.2
2.) Hardened concrete (age 28 days) :				
bulk density in kg/dm^3	2.27	2.35	2.36	2.31
compressive strength in N/mm^2	46.6	60.8	60.4	38.3

1) Degree of compactibility (*Verdichtungsmaß*) according to the German Standard DIN 1048
2) Flow diameter (*Ausbreitmaß*) according to the German Standard DIN 1048

Four test pieces for the slab method, four for the cube method, and five for the CDF test were produced for each concrete and each testing institute. The samples remained one day in the mould, and after de-moulding were then stored in accordance with the descriptions for the corresponding methods.

A total of 17 testing institutes took part in this Round Robin test, 15 used the Scandinavian slab method, 5 the German cube method and 6 the CDF test (see Table 3).

4   Presentation and discussion of the results

4.1   Resistance to the combination of frost and de-icing salt

Concretes 1 and 2A, which were identical in composition apart from the artificially introduced air voids in concrete 1, were tested for freeze-thaw resistance with de-icing salt. A 3% NaCl solution, which froze during the freezing process, was used for all the methods.

On the basis of the weathering losses all the institutes

were able to differentiate clearly between concrete 1 with high freeze-thaw resistance with de-icing salt and 2A with inadequate resistance (Fig. 4 and Table 4). For the slab and cube methods the average values for the components exposed to freezing increased virtually linearly with increasing number of freeze-thaw cycles (Figs. 5a + 5b), but with the CDF method (Fig. 5c) there was a progressive increase, especially for concrete 2A (up to 28 freeze-thaw cycles). This

Table 3   Participants of the 2nd European Round Robin Test "Freeze-thaw resistance of concrete"

No.	Country	Institute	German Cube Test	Scand. Slab Test	CDF - Test
01	Norway	NORCEM A/S	X	X	
02	Norway	NBI (BYGGFORSK)		X	X
03	Sweden	Statens Provings-anstalt	X	X	X
04	Denmark	Aalborg Portland A/S		X	
05	Finland	VTT Finland		X	X
06	UK	Rugby Cement		X	
07	France	VICAT	X	X	
08	Belgium	Centre National de Recherches Scientif. et Techniques		X	
09	Luxemburg	Laboratoire des Ponts et Chaussees		X	
10	Netherlands	CEMIJ (B.V.)		X	
11	Austria	Vereinigung der Österreichischen Zementfabrikanten		X	
12	Switzerland	EMPA		X	X
13	Italy	Italcementi S.P.A.	X	X	
14	Greece	Heracles General Cement Co.		X	
15	Germany	HAB Weimar			X
16	Germany	UNI-GHS-Essen			X
17	Germany	FIZ Düsseldorf	X	X	

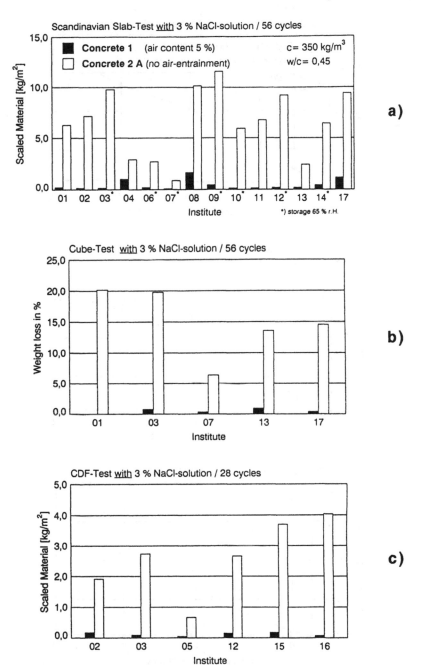

Fig. 4   Comparison between the results of the freeze-thaw
         tests with de-icing salt for all participants

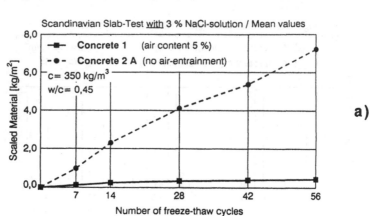

a)

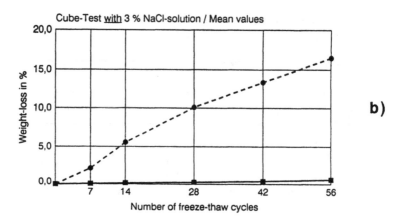

b)

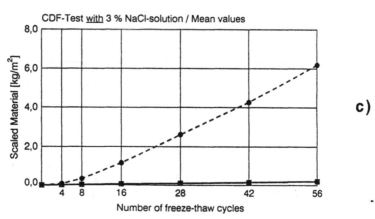

c)

Fig. 5   Mean values of scaled material and weight loss for the freeze-thaw tests with de-icing salt

method tests the moulded surface and it is possible that the initial freeze-thaw resistance with de-icing salt is slightly improved by the mould oil (release agent) remaining on the moulded surface.

With all methods the comparative variation in the results was relatively large. However, it is not possible to make a direct comparison between the methods as the cube test and the CDF method were each only carried out in a few institutes.

Scandinavian slab method
The weathering losses after 56 freeze-thaw cycles lay between 0.08 and 1.67 kg/m^2 for concrete 1, and between 0.88 and 11.6 kg/m^2 for concrete 2A. It was found that with the large variations there was some overlap. It is very noticeable that the weathering loss for concrete 1 was over 1.0 kg/m^2 at three institutes while for all other institutes it was less than 0.5 kg/m^2. The reason for this could not be explained. Perhaps the sample preparation, such as, for example, the cutting of the samples, had an effect on the result. It is not possible to detect any systematic influence of the differing preliminary storage conditions in this method (50% and 65% relative humidity in the climatic chamber).

Cube test
The weathering loss after 56 freeze-thaw cycles lay between 0.34 and 0.95 wt.% for concrete 1 and between 6.36 and 20.18 wt.% for concrete 2A. With the exception of institute 7 the results from the 5 institutes lay relatively close together. No explanation could be found for the sharply differing result from institute 7. The weathering loss measured in this institute was also very low for the Swedish slab method. Perhaps the air circulation is not intensive enough in the test chest used there.

CDF method
The weathering loss after 28 freeze-thaw cycles lay between 0.04 and 0.17 kg/m^2 for concrete 1 and between 0.66 and 4.03 kg/m^2 for concrete 2A. Four of the six institutes (institutes 3, 12, 15 and 16) had the same make of chest. Their results were relatively close together. Institutes 2 and 5 had chests which they had developed themselves. Their results - especially those from institute 5 - differed fairly sharply although the given temperature cycle was maintained. No explanation could be found for the differing results from institute 5.

Table 4   Ratios of mean values

Test method	Test with 3 % NaCl-solution				Test without de-icing salt		
concrete	1	2 A	**2A/1**	**2A/2B**	2 B	3	**3/2B**
Scand. Slab-Test (scaled material in kg/m^2) **56 cycles**	0.4	7.2	**16.4**	**72.3**	0.10	0.17	**1.7**
Cube-Test (Weight-loss in %) **56 cycles**	0.6	14.9	**24.0**	**62.0**	0.24	1.23	**5.1**
CDF-Test (scaled material in kg/m^2) **28 cycles**	0.1	2.6	**21.8**	**37.4**	0.07	0.24	**3.4**

4.2   Freeze-thaw resistance without de-icing salt

Concretes 2B and 3 were tested for freeze-thaw resistance. When using the cube method and the CDF test all the institutes were able to distinguish between concrete 2B with higher and concrete 3 with lower frost resistance (Table 4 and Figs. 6b, 6c, 7b and 7c). On the other hand, not all the institutes were able to make this differentiation when using the slab method (Fig. 6a). The difference between the two concretes may have been too small (Fig. 7a) and possibly, through testing the cut faces, there was also increased removal of aggregate particles which were not frost-resistant, which then falsified the results.

4.3   Dynamic elastic modulus

It had been expected that the results of the relative dynamic modulus would enable a clear differentiation to be made with respect to frost resistance of concrete. However, it was established that it was not yet possible to make a clear assessment. The research on this field is going on.

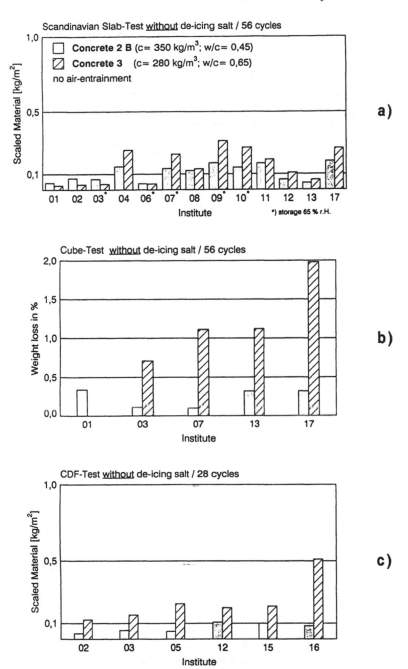

Fig. 6   Comparison between the results of the freeze-thaw
tests without de-icing salt for all participants

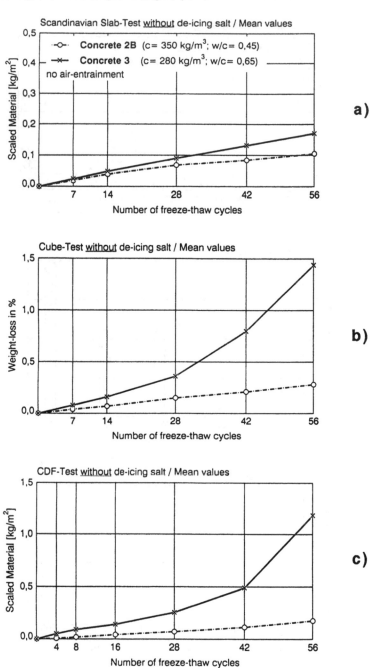

Fig. 7   Mean values of scaled material and weight loss for the freeze-thaw tests without de-icing salt

5 Summary and conclusion

With a suitable test method for determining the freeze-thaw resistance of concrete with and without de-icing salt, it must be possible to differentiate reliably between high, medium and low resistance. Any such method must be simple to perform and universally usable. The variation on repetition and the comparative variation should be as small as possible.

The following conclusions can be drawn from the European Round Robin test:

1) With all three of the methods used in the Round Robin test it was possible to differentiate between a concrete with high freeze-thaw resistance with de-icing salt and a concrete with inadequate resistance.

2) Testing the freeze-thaw resistance without de-icing salt showed that the weathering loss from the concrete with high freeze-thaw resistance (water/cement ratio = 0.45); cement content = 350 kg/m^3) was very low, and that of the concrete of lower freeze-thaw resistance (water/cement ratio = 0.65; cement content = 280 kg/m^3) was only slightly larger. It was possible to differentiate between the concretes somewhat better with the cube and CDF methods than with the slab method.

3) Measurement of the dynamic elastic modulus did not produce a clear differentiation between the concretes.

4) The Round Robin test emphasized the following advantages and disadvantages for the three methods:

   • Slab method
   The sample preparation is relatively complicated. The extent to which a cut face reflects practical behaviour during the test must be investigated. The test is simple to carry out. The costs of the test equipment and of carrying out the test are low.

   • Cube method
   Preparation and test procedure are simple. The costs of the test equipment and of carrying out the test are high.

- CDF method
  Preparation and test procedure are relatively simple.
  The temperature cycle can be maintained very accurately. The costs of the test equipment and of carrying out
  the test lie midway between those for the slab and cube
  tests.

In conclusion, it is recommended: firstly, that all three
methods be made more precise so that the comparative variations are reduced, and secondly, that all three methods be
included in the RILEM guidelines and further Round Robin
tests are carried out to find the precision datas.

6   Reference

1. Siebel, E.: Freeze-thaw-resistance of concrete with and
   without de-icing salt. beton 42 (1992) H.9, S. 496/501

2. Swedish Standard SS 137244. Concrete testing - Hardened
   concrete - Frost resistance.

3. Prüfung von Beton. Empfehlungen und Hinweise als Ergänzungen zu DIN 1048. Deutscher Ausschuß für Stahlbeton.
   Heft 422, Beuth-Verlag Berlin 1991.

4. Setzer, M.J., und V. Hartmann: CDF-Test Specifications.
   Betonwerk und Fertigteil-Technik 57 (1991), S. 83/86.

15

Mass loss experience with ASTM C 666: with and without deicing salt

D.J. JANSSEN
University of Washington, Seattle, WA, USA

M.B. SNYDER
University of Minnesota, Minneapolis, MN, USA

Abstract
Exposure to repeated cycles of freezing and thawing can lead to deterioration of portland cement concrete. The presence of deicing salts during the freezing and thawing is generally considered increase the severity of the exposure. A number of concrete mixtures having the same water-cementitious ratio (0.45) were exposed to repeated cycles of freezing and thawing in accordance with ASTM C 666. Deicing salt was included in some of the testing. For marginal concrete mixtures (durability factor approximately 35-65) the deicing salt had minimal effect on durability factor, but a significant effect on the amount of material lost through surface scaling. Inclusion of flyash or ground blast furnace slag in the mixtures generally reduced the amount of scaling.
Keywords: Durability factor, freezing and thawing, Philleo factor, salt scaling, spacing factor, specific surface.

1 Introduction

Damage to concrete from repeated cycles of freezing and thawing generally manifests itself in two forms: disintegration of the concrete mass and scaling of the concrete surface. In the past, disintegration was considered to be the more serious problem. Improvements in concrete (lower w/c, additions of pozzolans, and use of air-entraining admixtures) has for the most part eliminated occurrences of disintegration, while the increased use of deicing salts along with increased construction in areas exposed to seawater has led to an increased awareness of surface scaling as a significant deterioration phenomenon. This shift in perception of the predominant manifestation of frost-related deterioration has led to a need to re-examine test methods used to develop frost-resistant concrete mixtures.

Freeze-Thaw Durability of Concrete. Edited by J. Marchand, M. Pigeon and M. Setzer.
Published in 1997 by E & FN Spon, 2–6 Boundary Row, London SE1 8HN, UK.
ISBN 0 419 20000 2.

A standardized procedure for evaluating "Scaling Resistance of Concrete Surfaces Exposed to Deicing Chemicals", ASTM C 672, exists in the U.S. However, a recent survey of government agencies responsible for building and maintaining pavement systems found that almost none of the respondents performed ASTM C 672. Instead, ASTM C 666, "Resistance of Concrete to Rapid Freezing and Thawing" was the procedure most often performed. There are a number of possible explanations for this choice in test procedures, including the qualitative nature of the current ASTM scaling test as opposed to the quantitative nature of the rapid freezing and thawing test, the proliferation of equipment capable of performing rapid freezing and thawing, and the large experience base that many agencies have with rapid freezing and thawing.

Recognition that deicing salts can play a significant role in the deterioration of concrete has led some researchers to look at the behavior of concrete specimens subjected to repeated cycles of freezing and thawing (ASTM C 666) in salt water as opposed to fresh water. This variation on ASTM C 666 is not new, and in fact has been occurring for at least the last 25 years. [1] The purpose of this paper is to report on the results of a limited number of fairly conventional concrete mixtures, (w/(c+p) = 0.45) having marginal resistance to freezing and thawing, exposed to rapid freezing and thawing in both fresh and salt water. The intent of this limited testing was to provide guidance for future testing of the resistance of concrete to freezing and thawing in salt water. Comparisons will be made between durability factor (DF) values determined from specimens exposed to fresh and salt water, between mass loss and durability factor for specimens exposed to salt water, and between mass loss in salt water and various air-void parameters.

2 General observations

Specimens exposed to repeated cycles of freezing and thawing as prescribed in ASTM C 666 generally show little or no scaling as measured by either qualitative visual observation or quantitative mass loss measurements. This is true for both durable specimens and specimens that would be considered non-durable because their DF values were below 60. In terms of mass loss measurements, many non-durable specimens actually show a mass *increase* due to the absorption of water associated with internal deterioration. Mass increases of 0.1 to 0.2% are not uncommon, and mass loss often does not occur until isolated portions of the specimen fall off due to routine handling during testing. Exceptions to this trend of little or no scaling from testing in fresh water are specimens containing fairly high contents (4 to 5% by mass of cement) of calcium nitrite used as an accelerator. These specimens often show up to 1% or more mass loss due to scaling for either durable (DF > 60) or non-durable (DF < 60) mixtures. [2] In general, though, conventional mixtures do not scale significantly due to rapid freezing and thawing in fresh water.

3 Experimental program

3.1 Overview

To examine the influence of freezing and thawing in salt water (specifically a 3% by mass solution of NaCl) on behavior of marginal concrete mixtures, specimens from two 0.45 w/c mixtures containing marginal amounts of entrained air were subjected to freezing and thawing while exposed to either fresh water or salt water. Additional mixtures containing marginal air-void systems and various pozzolans as partial cement replacement were tested in salt water only. Relative dynamic modulus determined by fundamental transverse frequency and specimen mass were periodically measured. Air-void parameters including hardened air content, spacing factor $(\overline{L})$, specific surface, (α), and Philleo factor at 90% inclusion $(\overline{P}_{90})$ were determined for all mixtures.

3.2 Test specimens

Specimens were rectangular prisms with dimensions of 75 by 102 by 406 mm. This provided a surface area to volume ratio of approximately 0.05 mm^2/mm^3. The dimensions provided for measurement of fundamental transverse frequency about the 75 mm axis with minimal vibrational interference from either transverse vibrations across the 102 mm axis or torsional vibrations.

The mixtures tested all contained a vinsol resin air-entraining admixture and various combinations of ligno-sulfonate-based water-reducing admixture, Type F flyash, Type C flyash, or ground blast furnace slag (GBFS). All had a water-cementitious ratio (w/(c+p)) by mass of 0.45 and contained the same coarse and fine aggregates. The mixtures were low slump (25-50 mm) and attempted to duplicate mixtures typical for concrete pavements. Paste contents were in the range of 25 to 27% of the volume of the concrete. The mixtures are summarized in Table 1. Mixtures N26 and N27 were cured 14 days in a lime-water bath while the remaining mixtures were cured for 28 days in lime-water. All mixtures were prepared at low air contents in order to produce mixtures with marginal DF values.

Table 1. Summary of test mixtures

Mixture number[*]	Water reducer	Pozzolan
N26	None	None
N27	None	None
N51	None	None
N52	None	None
N53	Yes	None
N54	Yes	None
D10	Yes	15% Class F
D18	Yes	15% Class C
D30	Yes	40% GBFS

[*] These mixtures were part of a larger study. [2]

3.3 Test procedures
All specimens were subjected to repeated cycles of freezing and thawing as described in ASTM C 666; that is the temperature measured in the center of a control specimen cooled from 5°C to -18°C, and then warmed back to 5°C in approximately four hours. The heat-transfer medium was refrigerated air during the cooling portion of the cycle and water at 6°C during the warming portion. Specimens were tested in a vertical position. The specimen conditions of either wet with fresh water or wet with salt water were maintained as follows:

- Fresh Water - Specimens were wrapped with cotton terry-cloth. This cloth absorbed water during the warming portion of the cycle and retained the moisture during the freezing portion. The use of the cloth wrap eliminated external forces on the specimens associated with the use of water-filled containers. These specimens are referred to as "wrap" in the results and analysis.
- Salt Water - Containers were necessary for testing in the 3% NaCl solution. The containers were made from two pieces of vacuum-formed plastic which were glued together to form a container providing the one- to three-mm space surrounding the specimens as described in ASTM C 666. Each specimen container was emptied and refilled to approximately three mm above the top of the specimen with 3% NaCl solution whenever specimens were tested for fundamental transverse frequency and mass loss.

Periodic measurement of fundamental transverse frequency and specimen mass were made to the specimens until the specimens had either experienced at least 300 cycles of freezing and thawing or until the relative dynamic modulus of the specimens dropped below 60%. Relative dynamic calculations were corrected to consider mass loss for specimens tested in salt water.

3.4 Linear traverse
Hardened air-void parameters were determined by examining specimen areas of 7,750 mm^2 as described in ASTM C 457. The total traverse length for each specimen was 2.3 m. Hardened air content, $\bar{L}$, and α were determined by standard procedures. Individual air-void chord lengths were recorded so that a best-fit zero-order log-normal distribution for those chords under 1.0 mm in length could be determined. [3] This distribution was then used in procedures described by Lord and Willis [4] and Philleo [5] to determine $\bar{P}_{90}$. This value is the maximum distance from an air void for 90% of the paste. While no specific criteria for maximum $\bar{P}_{90}$ value for durable concrete has been established, examination of a number of specimens [2] with $\bar{L}$ values of 0.20 mm and a values of 25 mm^2/mm^3 suggests that a $\bar{P}_{90}$ value of 0.040 mm would be appropriate.

4 Discussion of results

4.1 Comparison of testing in fresh water and salt water
Initial testing consisted of mixtures N26 and N27, with five specimens from each mixture being tested in wraps and five specimens of each being tested in 3% NaCl

solution. Table 2 shows freezing and thawing and linear traverse results for these mixtures.

Table 2. Results of 14-day cure mixtures in wraps and 3% NaCl

Mixture number	Test condition	Durability factor (DF)	Mass loss %	Cycles to 5% mass loss
N26	Wrap	36	none	NA
	Salt	45	14.4	47
N27	Wrap	66	none	NA
	Salt	64	12.8	67

A closer examination of the changes occurring with repeated cycles of freezing and thawing is shown in Figures 1 and 3 for mixture N26 and Figures 2 and 4 for mixture N27. Figures 1 and 2 show mass loss for the specimens tested in salt water. Mixtures tested in fresh water (wraps) had no mass loss. Figures 3 and 4 present change in relative dynamic modulus, corrected for mass loss. For these mixtures having marginal durability, the use of salt water instead of fresh water has only a minimal effect on change in relative dynamic modulus. This is probably due to the outside-to-inside nature of deterioration in specimens subjected to repeated freezing and thawing. In salt water, the deteriorated material on the outside of the specimens merely scales off.

4.2 Comparison of durability factor and mass loss

Table 3 summarizes the DF, mass loss at the end of ASTM C 666, and cycles of freezing and thawing to produce a mass loss of 5%. These results along with values from Table 2 for the 14-day cure specimens tested in salt water are presented graphically in Figures 5A and B. Figure 5A shows the percent mass loss when the specimens reached 60% relative dynamic modulus (or 300 cycles of freezing and

Table 3. Results of 28-day cure specimens tested in 3% NaCl

Mixture number	Pozzolan and/or WR	Durability factor (DF)	Mass loss %	Cycles to 5% mass loss
N51	None	41	11.0	79
N52		33	14.9	42
N53	WR	38	10.3	70
N54		31	11.0	58
D10	WR, 15% F	37	8.9	142
D18	WR, 15% C	21	9.3	68
D30	WR, 40% GBFS	31	6.3	131

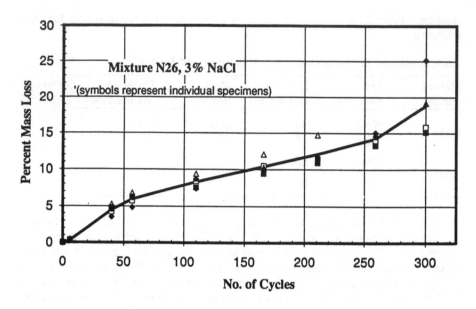

Figure 1. Mass loss versus cycles of freezing and thawing, mixture N26.

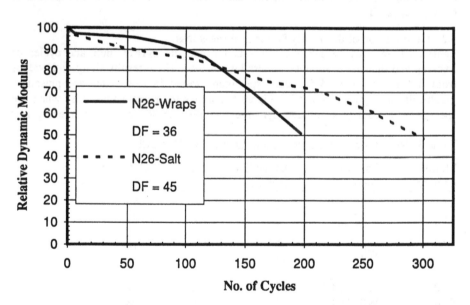

Figure 2. Relative dynamic modulus versus cycles of freezing and thawing, mixture N26.

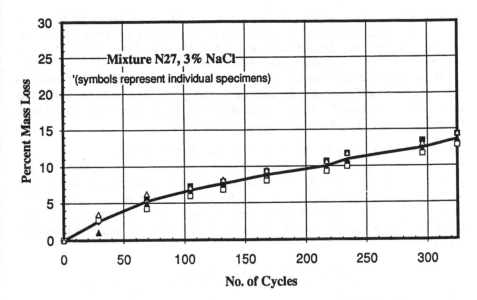

Figure 3. Mass loss versus cycles of freezing and thawing, mixture N27.

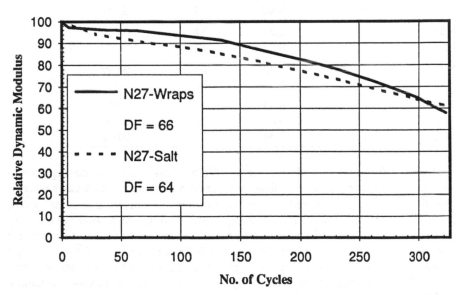

Figure 4. Relative dynamic modulus versus cycles of freezing and thawing, mixture N27.

thawing for the mixture with a DF value greater than 60), while Figure 5B shows the number of freezing and thawing cycles required to produce 5% mass loss for the various mixtures. For the specimen size tested, 5% mass loss corresponds to an average depth of scaling of approximately one mm.

These figures clearly support the generally-accepted notion that the correlation between DF and surface scaling in salt water is very poor. In some cases, scaling can be substantially reduced by partial replacement of cement by a pozzolan even though the mixture may still have a low DF value.

4.3 Effect of air-void parameters on durability factor and mass loss

Air-void parameters determined from hardened specimens for the various mixtures are summarized in Table 4. Though other parameters are presented, $\bar{P}_{90}$ will be used for further analysis because it is the single parameter that best characterizes the protection provided to the paste by the entrained air. This is because $\bar{P}_{90}$ quantifies the distance in the paste to an air void without the limiting assumptions associated with $\bar{L}$. [5]

Table 4. Air-void parameters determined by linear traverse

Mixture number	Hardened air %	$\bar{L}$ mm	α mm²/mm³	$\bar{P}_{90}$ mm
N26	2.4	0.37	18	0.086
N27	2.8	0.40	15	0.080
N51	2.0	0.27	27	0.079
N52	1.5	0.37	22	0.099
N53	1.6	0.33	23	0.099
N54	1.5	0.49	16	0.131
D10	3.1	0.33	22	0.070
D18	1.3	0.40	26	0.105
D30	3.0	0.44	17	0.092

Figure 6 shows DF plotted versus $\bar{P}_{90}$. While at first glance the data appears to only represent scatter, closer examination suggests some trends. In all cases where a specific mixture was duplicated at a slightly different air content, a decrease in $\bar{P}_{90}$ results in an increase in DF. This is as expected. Also, use of a water-reducer may result in improvement in DF. In all cases, replacement of a portion of the cement with pozzolan resulted in a slight decrease in DF for similar values of $\bar{P}_{90}$. Though all mixtures shown have the same w/(c+p), the actual w/c of the pozzolan mixtures was 0.52 for the mixtures containing flyash and 0.75 for the mixture containing GBFS.

Cycles of freezing and thawing in salt water to produce 5% mass loss is shown in Figure 7 as a function of $\bar{P}_{90}$. Again, use of a water-reducer improved performance by increasing the number of cycles necessary to produce a mass loss of 5%. However, the use of a pozzolan to replace cement resulted in improved performance with respect to mass loss. This agrees with general observations that field

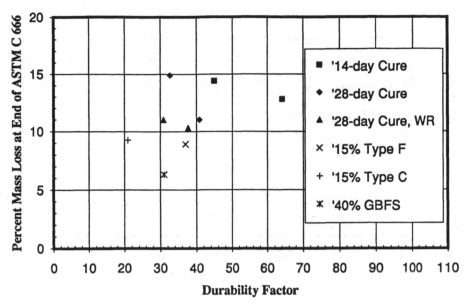

Figure 5A. Mass loss at end of ASTM C 666 versus durability factor, 0.45 w/(c+p) mixtures in 3% NaCl solution.

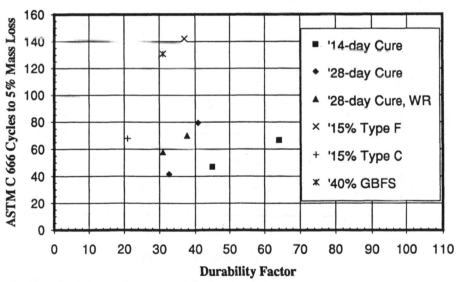

Figure 5B. Cycles of freezing and thawing to 5% mass loss versus durability factor, 0.45 w/(c+p) mixtures in 3% NaCl solution.

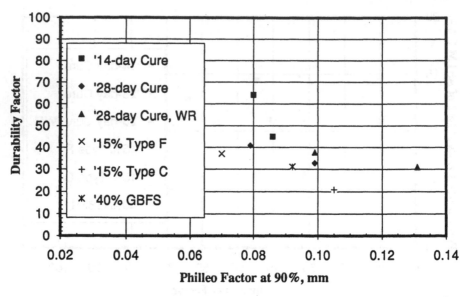

Figure 6. Durability factor versus Philleo factor at 90%, 0.45 w/(c+p) mixtures in 3% NaCl solution.

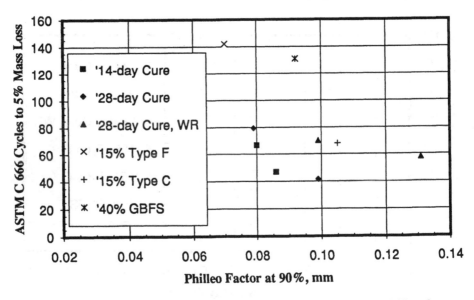

Figure 7. Cycles of freezing and thawing to 5% mass loss versus Philleo factor, 0.45 w/(c+p) mixtures in 3% NaCl solution.

performance is enhanced by judicious use of pozzolans as partial cement replacements. The conflict between DF and scaling performance for mixtures containing pozzolans supports the opinion that the test method used and measurements taken should reflect the anticipated damage manifestation: mass disintegration or surface scaling.

5 Conclusions

While the scope of the testing reported in this paper is far too limited to permit specific conclusions concerning the resistance to damage from freezing and thawing in salt water of concrete mixtures made with and without water-reducing admixtures and/or pozzolans, some general conclusions appear to be supported:

1. Testing by ASTM C 666 in 3% NaCl may not produce results significantly different than those produced by testing in fresh water when only relative dynamic modulus measurements are considered.
2. No clear relationship appears to exist between DF and mass loss for specimens tested in 3% NaCl.
3. The use of some pozzolans as cement replacements can improve the resistance of concrete to scaling in 3% NaCl, while possibly having a slightly negative effect on DF values determined by testing in 3% NaCl.
4. Frost-resistance criteria should be developed with consideration to the expected exposure conditions of the concrete. Specifically, the criteria for concrete exposed to salt water from deicing salts or other sources should consider the scaling potential of the mixture.

6 Acknowledgments

The authors wish to thank the Strategic Highway Research Program and the National Science Foundation for their support of this work. This research was conducted at the University of Washington and Michigan State University.

7 References

1. Newlon, H.H. (1978) *Modification of ASTM C 666 for Testing Resistance of Concrete to Freezing and Thawing in Sodium Chloride Solution*, Virginia Highway & Transportation Research Council, Charlottesville, VA VHTRC 79-R16.
2. Janssen, D.J., and Snyder, M.B. (1994) *Resistance of Concrete to Freezing and Thawing*, SHRP-C-391, Transportation Research Board.
3. Roberts, L.R., and Scheiner, P. (1981) Air void system and frost resistance of concrete containing superplasticizers. *Developments in the Use of Superplasticizers*, ACI-SP-81.

4.   Lord, G.W. and Willis, T.F. (1951) Calculation of air bubble size distribution from results of Rosiwal traverse of aerated concrete. ASTM Bulletin 177, pp. 56-61.

5.   Philleo, R.E. (1955) A method for analyzing void distribution in air-entrained concrete. Portland Cement Association Research and Development Division.

16

Internal comparative tests on frost-deicing-salt resistance

W. STUDER
EMPA, Dübendorf, Switzerland

Abstract
This paper deals with the influence of the type of temperature cycle on the amount of scaled off material in rapid weathering procedures to assess the frost-deicing-salt resistance of concrete. It is based on results of internal comparative tests.

It was found that the severeness of the temperature cycles can be well characterized by the minimum temperature reached in the salt solution layer on top of the specimens, but that the relation between temperature cycle and damage is strongly non-linear and depends on some other important factors as e.g. the saturation conditions.

Out of this, some conclusions on the test procedure are drawn.

Keywords: Frost-deicing salt resistance, minimum temperature, saturation conditions, temperature cycle.

1 Introduction

The actual reason for conducting the internal comparative tests on frost-deicing-salt resistance described here was the fact, that our freezing cabinets are approaching the end of their lifetime and will soon need to be replaced.

We have therefore purchased a Swedish apparatus that was developed for the so-called Borås Test [1], and which has proved satisfactory, in order to clarify the possibility of using it to carry out the Swiss frost-deicing-salt resistance test according to SIA 162/1 [2,3].

The two procedures differ from each other essentially in the temperature cycle, the manner of pre-storage and the thermal insulation of the test specimens during the tests. The internal comparative tests were thus also able to be used to investigate the influence of these parameters and thus establish a relationship between the SIA and Borås Tests, on the one hand, and to complement the European Round-Robin Tests on the other hand.

In the present contribution, the emphasis is on the influence of the temperature cycle.

Freeze-Thaw Durability of Concrete. Edited by J. Marchand, M. Pigeon and M. Setzer.
Published in 1997 by E & FN Spon, 2–6 Boundary Row, London SE1 8HN, UK.
ISBN 0 419 20000 2.

2 Scope of investigation

The combinations of parameters selected for the tests are summarized in Table 1. As already mentioned, the main goal was a direct comparison between the old and the new cabinet, which was the reason for conducting Series 11...13 and 21...24 simultaneously. For the comparison of Series 2., 3., and 4., the parameter "concrete age" could not be kept constant for technical reasons (only one new cabinet). A repetition of Series 1. and 2., however, allows one to estimate the influence of this parameter.

Table 1   Combinations of parameters

Series No.	11...13	21...24	31...32	41...44
Cabinet	old	new	new	new
Temp. cycle: Duration Max/Min Temp.	16.8h +13/-13 °C	12 h +13/-13 °C	24 h +20/-20 °C	12 h +20/-20 °C
Thermal insulation	no	no	yes	no
Pre-storage Duration	7 d	7 d	3 d	3 d/7 d

Series 11 :   SN 565 162/1
Series 32 :   SS 13 72 44

Three types of concrete were examined, which in our experience show high, medium and low frost-deicing-salt resistance, respectively. Table 2 summarizes the properties of the fresh and hardened concrete.

For each type of concrete, a batch of 0.15 m^3 was prepared and used to cast 17 or 18 cubes of 200/200/200 mm.

From these cubes, 150/150/50 mm test slabs were taken, the test surfaces being either lateral mould surfaces or cut surfaces parallel to these. The test surfaces were treated according to the relevant instructions for testing (adhesion of rubber/foam rubber, thermal insulation).

This work was carried out at various ages of concrete and at different times before the start of testing. All intermediate storage, however, was at 20 °C and 70 % RH.

At the start of testing, i.e. the start of frost cycling, the age of the concrete was 109...118 d for Series 1 and 2, 193...200 d for Series 3,137...144 d for Series 4 and 165...175 d for the repetition of Series 1 and 2.

During frost cycling, an approx. 3 mm layer of 3% NaCl solution was present on all test surfaces.

The parameter measured and used for comparison was the dry mass of scaled off material after certain numbers of frost cycles.

Table 2. Composition and properties of the concrete

Type of concrete	1	2	3
Composition [kg/m³]:			
Sand 0/4 mm	639	716	730
Gravel 4/32 mm	1347	1168	1192
Cement PC	300	293	299
Air entraining agent	--	1.5	--
Water	150	146	179
Fresh concrete properties			
Apparent density [kg/m³]	2446	2325	2400
Air pores content [vol.%]	1.7	5.2	2.1
W/C-ratio (index)	0.5	0.5	0.6
Degree of compactibility	1.16	1.16	1.03
Degree of segregability	1.43	2.45	2.11
Hardened concrete properties:			
(age: approx. 120d)			
Compressive strength [N/mm²]	52.6	41.5	39.9
Fraction of pores fillable by capillary suction [vol.%]			
Surface layer	13.16	15.06	17.17
Inside	10.91	11.58	12.19
Fraction of air pores [vol.%]			
Surface layer	1.42	5.65	1.81
Inside	1.23	3.92	1.64
FS			
Surface layer	1.3	1.9	1.2
Inside	1.4	1.9	1.4

Degree of segregability: ratio of fine mortar content in
    top layer to content in bottom layer after 60s
    vibration in degree of compactibility measuring vessel
    (according to Walz)
FS: ratio of air pore content to free pore volume at cri-
    tical degree of saturation = measure of frost resi-
    stance

    FS ≥ 1.5 :   high frost resistance
    1.0 < FS < 1.5 :   medium frost resistance
    FS ≤ 1.0   :   low frost resistance

Apart from this, the changes in mass of the test specimens
were measured.
    In separate tests, the curves of the temperature in the
salt solution layer against time were determined for
various locations in the cabinets and different numbers of
specimens. The respective curves may be regarded as
characteristic for to the individual test specimens
according to their positions during frost testing.

Table 3. Means of series and standard deviation (± S)of the amount of scaled off material [g/m²] at 30 frost cycles (Series 11...13) and 28 cycles (all others), respectively

Series	Concrete 1		Concrete 2		Concrete 3	
	Mean	± S	Mean	± S	Mean	± S
11	443	337	33	23	2087	1713
12	1577	448	110	35	2786	1397
13	1046	1007	15	8	1430	713
21	602	410	96	31	903	348
22	1186	1135	69	40	3828	1931
23	479	273	73	37	768	525
24	952	19	11	7	2421	1603
41	2668	340	162	153	5075	895
42	3523	1337	99	21	5302	815
43	1809	927	150	94	4098	1627
44	2335	713	101	39	4387	2498
11 W	1343	275	32	22	3132	1605
12 W	2973	1748	89	31	3751	537
13 W	2889	3317	15	7	1009	816
21 W	1303	268	37	28	2055	890
22 W	1706	139	53	22	2991	1086
24 W	866	403	26	15	3070	3977
31	5819	1321	275	183	8391	4055
32	3525	1275	61	36	4385	1377

*W = Repetition of series

Table 4   Freezing and thawing cycle characteristics Temperature in the salt solution layer.

	Minimum temperature [°C]			Freezing rate [°C/h]		
	Mean	Max	Min	Mean	Max	Min
11...13	-13	-13	-13	-3.38	-4.27	-2.91
21...24	-11	-14	-7	-2.33	-4.07	-1.25
31...32	-18	-19	-17	-2.03	-2.39	-1.63
41...44	-16	-20	-12	-3.42	-5.00	-1.94

3 Results

Table 3 summarizes the means and standard deviations of the amounts of scaled off material after 28 (30) frost cycles for the individual series.

The half-time of the test was selected because up to this point in time, the salt solution layer was always fully present on all specimens. Later on, the layer was sometimes lost between measurements with the poor quality concrete types. This caused the amounts of scaled off material to be drastically reduced and the measurements to be rejected.

Since the progress of damage was essentially similar for all specimens of the same concrete (cf. Fig.1), the "instantaneous examination" at the half-time of the runs may be considered as genuinely characteristic.

In Figure 2, the mean and the extreme temperature cycles in the salt solution layer are shown for the groups of series 1. to 4..In Table 4, the minimum temperature in the solution layer and the freezing rate (= mean speed of temperature drop from 0 °C to -5 °C) are sumarized. These values are considered to be the main characteristies of the frost cycles.

Figure 3 shows that although the minimum temperature has a clear effect on the amount of scaled off material, the scatter between specimens is, on the other hand, so great that this influence is not detectable within the series. For this reason, the further discussion will only consider the means of series, both for the amounts of scaled off material and for the temperature-time curves.

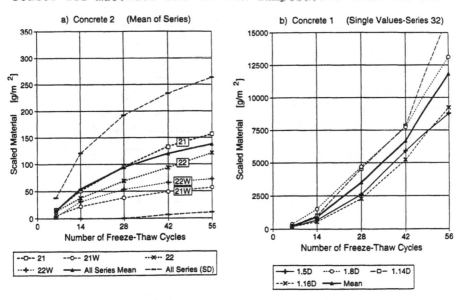

Fig.1. Progress of damage: Similarity between and within series

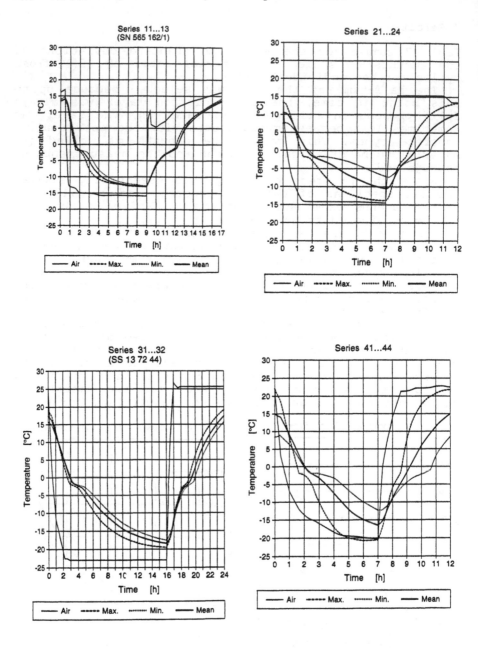

Fig.2. Mean and extreme temperature cycles

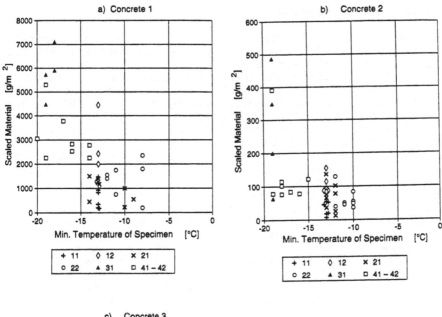

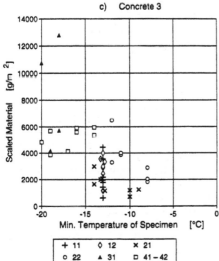

Fig.3. Amounts of scaled off material (individual specimens) after 28 frost cycles as a function of the minimum temperature in the salt solution layer.

4 Discussion of results

With one exception (Series 13 W: cut surface, old cabinet, +13/-13 °C in 16.8 h), the three types of concrete had the same ranking in all series:
The air entrained concrete (Concrete 2: W/C ratio = 0.5, air content = 5.2 vol.%), as expected, showed the best frost-deicing-salt resistance by far. The EMPA Standard Concrete (Concrete 1: W/C ratio = 0.5, air content = 1.7 vol.%) took the second place and the sand-rich concrete (Concrete 3: W/C ratio = 0.6, air content = 2.1 vol.%) had the lowest frost resistance. On average, the amounts of scaled off material after 28 or 30 cycles were in the ratio of approximately 1 (Concrete 2) : 20 (Concrete 1) : 40 (Concrete 3).
The same ranking results also from the porosity of the surface layer (cf. Table 2) with FS = 1.9 (Concrete 2), FS = 1.3 (Concrete 1) and FS = 1.2 (Concrete 3).
On repeating Series 11, 12 and 13 as well as 21, 22 and 24, a tendency to larger amounts of detached material was found. Because of the small series, the large scattering and the partly "unknown" history of the test specimens, it is not admissible, however, to deduce that the frost-deicing-salt resistance decreases with age.
For the assessment of the influence of the temperature cycle, Series 11, 12, 21, 22, 31, 41 and 42 were considered, in which the mould surface was tested and prestorage with water was carried out. For Series 11...22, the mean of series and repetition was used in each case. In Figure 4, the mean relative amount of scaled off material (referred to Series 31) is plotted as a function of the mean minimum temperature in the salt solution layer after 28 or 30 frost cycles (values from Tables 3 and 4).
As may be seen, there exists for all three types of concrete the same strongly non-linear relationship between the measured values. The minimum temperature thus appears to be a good characterization of the "severity" of the frost loading.
The other measure of intensity selected in Table 4, the "freezing rate", is less suited for this purpose.
Furthermore, it turns out that variations of the minimum temperature in the range of -18 °C (Borås Test) have a considerably greater effect than at -13 °C (SIA Test):
A "reduction" of the minimum temperature from - 18°C to -16 °C causes a drop in the amount of scaled off material by 38...52% as compared with only 4...22% for the reduction from -13 °C to -11 °C.
To reach the same scattering of results , the temperature control must therefore be more exact, the lower the minimum temperature is chosen. This causes problems with air-cooled cabinets, which are critical in this respect.
The choice of a moderate minimum temperature in the range of -10 °C...-15 °C is also supported by the fact

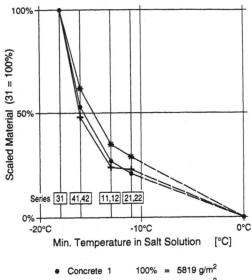

Fig.4. Relative amount of scaled off material as a
function of the minium temperature in the salt
solution layer (mean values)

that down to this value, the assumption of a linear effect
on the amount of scaled off material appears reasonable
and hence an extrapolation of the test results to the
behaviour of structural concrete under natural frost cycling (see Fig.5) becomes more reliable.
In spite of this, the relationship between testing and
practice remains rather diffuse.
For example, it has turned out that practically no scaled material arises when the salt solution layer is mis-
sing or is replaced by a water layer during frost cycling.
These observations, by the way, have not yet been able to
be explained satisfactorily and are the subject of further
investigations.
These evidently important "saturation conditions" are,
to a large extent, unknown during natural frost cycles. It
can only be conjectured that they are considerably more
favourable than in the tests and that for this reason the
test results lie considerably on the safe side in this
respect. However, this is of little help, since the number
of frost cycles to which structural concrete is subjected
during its lifetime is 50...100 times greater than in the
tests.
Finally, one must consider that in general, only such
concrete is tested that should exhibit a high frost-

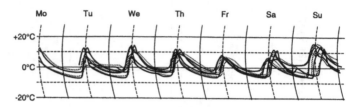

Fig. 5. Natural temperature cycles (alt.: 900 m above sea
level, Jan.-Febr. 1989)

deicing-salt resistance, i.e. a maximum of a few $100g/m^2$
scaled material at the end of the test period or lifetime.
The concrete layer observed is hence only a fraction of a
mm thick and it is hardly possible at all to measure or
even to control the relevant properties of this layer,
assuming they are even known. It suffices here to refer to
the influence of the content and arrangement of unstable
aggregate particles.

5 Summary and conclusions

Internal comparative tests were conducted to clarify
whether the frost cabinets at the EMPA for the frost
resistance test according to SIA 162/1 can be replaced by
much less expensive and easy-to-operate frost cabinets of
Swedish origin. During these tests, the influence of the
temperature cycle on the extent of damage was able to be
determined. For this purpose 67 test specimens, each
150/150/50 mm in size, were prepared from three types of
concrete showing, according to experience, high, medium
and low frost-deicing-salt resistance, respectively, and
tested in 13 series (of which 6 were repeated). The main
parameter varied was the temperature cycle, measured in
the salt solution layer. As secondary parameters, the pre-
storage, the preparation of the test specimens and the
type of the test surface were varied.
   The dry mass of scaled off material after certain numbers of frost cycles served as criterion quantity for
measurement and comparison.
The following conclusions can be drawn from the results of
the tests:

- The temperature cycle in the salt solution layer has a
  decisive influence on the amount of scaled off material
  and may be well characterized by the minimum temperature reached.
- The duration of the frost cycle, the freezing rate and
  the temperature-time curve in the thawing phase didn't
  show a discernible influence.

- The evolution of scaling during frost cycling was similar for all series of one concrete type, practically independent of the final amount of scaled off material. In particular, the distinction between the three types of concrete was not clearer at minimum temperatures of -16...-18 °C than at -11...-13 °C and not better at 56 frost cycles than at 28 cycles.

- The influence of the minimum temperature reached during frost cycles on the extent of damage increases with lower temperatures. Due to the change from -18 °C to -16 °C the amount of scaled off material was reduced by 38...52% as compared with only 4...22% reduction due to the change from -13 °C to -11 °C.
  Extrapolation of test results to the behaviour of the concrete under natural frost cycling, where minimum temperatures mostly are above -5 °C, seems therefore to be more reliable with moderate (~ -10 °C) than with low minimum temperatures (~ -20 °C).

- The relation between the behaviour of concrete in the test and under natural frost cycling is rather weak, since the "saturation conditions" for natural frost cycling are to a large extent unknown, testing concerns a concrete layer of at most one millimetre in thickness, the scatter of the relevant material properties is large and 50 to 100 times more frost cycles occur on the building during its lifetime than in testing. Thus this kind of tests allow, at most, a statement of the kind: "Small amounts of scaled off material on test specimens result in a low probability of serious damage on a building".

- The effort for a rapid weathering procedure for the assessment of frost-deicing-salt resistance can therefore, in our opinion, be kept at a low level without significant loss of information by selecting a moderate minimum temperature (e.g. -12 °C), a low number of frost cycles (e.g. 28) and a simple specimen preparation (e.g. adhesion of a rubber band, prestorage under water).
  For research on the damage mechanism, however, such a simple procedure is hardly suitable.

- The guarantee of a high frost-deicing-salt resistance on structural concrete cannot be served by a rapid weathering test procedure - except, perhaps, through the psychological pressure that can be exerted by an inspection and associated sanctions. For that purpose, a control of fresh concrete is necessary, which would allow unsatisfactory concrete mixtures to be recognized and rejected before casting.
  That this is actually possible is shown by our experience with the method of the critical degree of

saturation for the determination of the frost resistance [4].

6 References

1. Swedish Standard SS 13 72 44 (March 1992) Concrete testing Hardened Concrete - Frost resistance

2. Swiss Standard SN 562 162/1 (1989) Betonbauten Materialprüfung

3. EMPA Guidelines for testing (1989) Testing of water conductivity, frostresistance and frost-thawing salt resistance, EMPA, Dübendorf, Switzerland.

4. Studer, W. (1984) Testing of Frost Resistance on Fresh Concrete, Proc. RILEM Seminar: Durability of concrete structures under Normal Outdoor Exposure, 26th - 29th March 1984, Hannover, Institut für Baustoffkunde und Materialprüfung Universität Hannover, Germany.

Author index

Subject Index

9 780367 863999